# PatenTRAIL

# DIY Path to Securing a Patent for $800 or Less

Roy Lique

(This book is also available in Kindle (mobi) format at Amazon)

# Table of Contents

## Table of Contents

Table of Contents

# PART 6 – PROSECUTION OF APPLICATION ... 207

# Book Title Validation

The total cost of securing the patent to Digital Camera Lens Guard and Use Extender is amazingly low that it needs validating, lest the book becomes the subject of doubtful reception.

In summary, USPTO's charges include: $125 for filing the Provisional Patent Application, $435 for filing and prosecution of the Non-Provisional Patent Application, and $240 for the issuance of the Patent, for a total cost of $800.

Were I to do an application for a patent again, depending on the competitive nature of the invention, I would consider avoiding filing the Provisional Application thereby saving me $125.

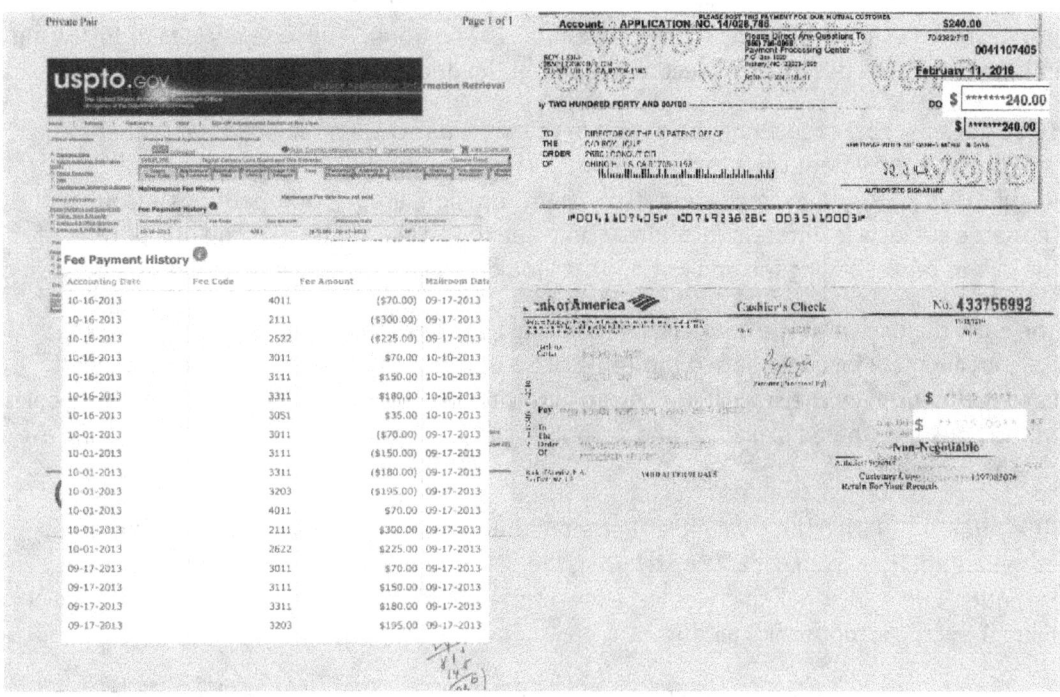

For the services the USPTO renders, $675 would be an attractive investment on an idea that could potentially return financial rewards. On the other hand if the patent application is denied for any reason, the loss is not too material.

# Preface

Writing a book on any subject was never in my mind from the beginning, not to mention a book about invention. Obstacles exist in every facet of writing, particularly in my case.

As I realistically found out, writing a book about invention needs a significant amount of precise understanding of the English language. English is not my native language forcing me to be more careful about the use of words to communicate and project my ideas to the readers including patent examiners, resulting in longer periods to finish a project. To boot, my study of the English language was limited to the high school level that English was taught in a foreign country.

This book is written to demonstrate how I, with very limited resources, am able to secure a real patent on the invention called Digital Camera Lens Guard and Use Extender. The book is not designed to teach how to secure a patent on an invention. If anyone decides to follow and imitate the steps I described to secure the patent, he does it on his own risk and volition.

To achieve my goal of securing the patent in the most direct way, I skipped the sections and provisions in the Patent Laws that did not apply to my purpose. The Code of Federal Regulations (CFR) and its implementations, in my opinion, are full of repetitious accounts of the same subjects in different locations. I only needed a clear understanding of a portion of the law to proceed filing the application that would result in the issuance of the patent.

The early major part of the book cites, in the exact words of the code, the provisions in the Patent Law that I followed. I used the Table of Contents of the Law as a road map to tightly keep me focused on the subject of filing an accurate application for the patent. The later parts of the book are where I explained and described the procedures that resulted in securing the patent.

As a reference, I used the invaluable "Patent It Yourself", by Patent Attorney David Pressman, 16th Edition, from Nolo.com. After securing the patent on Digital Camera Lens Guard and Use Extender, I honestly think that would-be inventors should be wary about whom they entrust filing their applications for a patent.

As I found out, in my opinion, patentability of a product is not decided until an application has gone through the examination process. By then an enormous amount of money and time would have been spent on the application. DIYing an application for a patent is the way I would go through if I have to apply for one again.

# Acknowledgements

My special thanks to the staff of the United States Patent and Trademark Office, Art Unit 2852, particularly Primary Examiner Christopher E. Mahoney, for the professional and expeditious prosecution of the application resulting in the grant of the Patent presented in this book.

I also want to acknowledge the efficiency of the USPTO staff for ably handling the Patent Application Information Retrieval (PAIR) system which saves applicants money in reduced filing fees and time in terms of accessing organized files relating to applications.

I further want to acknowledge the invaluable "Patent It Yourself", 16th Edition, by Patent Attorney David Pressman, from Nolo Press, as a resource without which I would have never been able to prepare an application for the patent presented in this book.

# The Birth of an Idea and the need to invent

Necessity is the mother of invention, so it has been said and known all the time. Suggesting urgency, invention implies serious dedication of time, money, and efforts to the project. There are, however, other motivations for inventing - leisure is one.

We were at Amalie in the Caribbean taking a leisurely cruise. Entertainments of many forms naturally awaited us at the ports and on the shores of each destination. Since we could not make it to all the locations of entertainments without being tired, we just looked and listened from afar what were going on in certain locations. That bothered me having to take blurry pictures from a distance. An idea was starting to form.

My friend was carrying a camera and a bagful of lenses. I could see him shifting the bag from his left hand to the right hand and vice versa, an indication that the weight of the photographic equipments was negatively affecting his ability to completely enjoy the cruise vacation. On the other hand, I carried a small camcorder which was considerably lighter but still limited in terms of taking pictures from a distance. A digital camera would be an ideal size for traveling lightly, so I was thinking, but it still had the same limitation as the camcorder.

What would be the result, I wondered, if a digital camera is fitted with a sighting device such as a monocular and have it take a picture of what the monocular sees. A monocular can focus on a subject from afar but it cannot take pictures. A digital camera on the other hand, can take clear pictures of close objects but poor quality ones of objects taken from a distance. What, I wondered further, if a construction of coupling a camera and a monocular is made for possible financial and leisurely rewards. This brought up the motivation for inventing the Digital Camera Lens Guard and Use Extender.

The financial rewards aspect of inventing was even more motivating when at the back of certain forms relating to patents, I found the following:

### SelectUSA
The United States represents the largest, most dynamic marketplace in the world and is an unparalleled location for business investment, innovation, and commercialization of new technologies. The U.S. offers tremendous resources and advantages for those who invest and manufacture goods here. Through SelectUSA, our nation works to promote and facilitate business investment. SelectUSA provides information assistance to the international investor community; serves as an ombudsman for existing and potential investors; advocates on behalf of U.S. cities, states, and regions competing for global investment; and counsels U.S. economic development organizations on investment attraction best practices. To learn more about why the United States is the best country in the world to develop technology, manufacture products, deliver services, and grow your business, visit http://www.SelectUSA.gov or call +1-202-482-6800.

## The Birth of an Idea and the need to invent

With the above information and motivations, the idea about The Digital Camera Lens Guard and Use Extender was born. How the process of securing the patent for the idea is the object of writing this book in the most accurate, thorough, and inexpensive manner –  $800 DIY.

# Introduction

With the complexity of the Patent Laws and with no knowledge of how to start the application for a patent, I had to devise a plan of action to simplify my approach. Accordingly, I developed a rough and simplified outline and a roadmap in the form of a table of contents that I obtained from the website of the United States Patent and Trademarks Office, to help me focus on the project. By using the outline and the roadmap, I avoided the parts of the law that I did not need, referring to them only when absolutely necessary.

Following the outline and the roadmap, I initially show the actual certificate granting me the patent rights to the Digital Camera Lens Guard and Use Extender. The outline and roadmap served as guides in the development of the whole book, the goal in this literary exercise.

### The goal - actual patent as object of development of this book

Start with an idea
> How an idea became the object of an invention
> Prototyping to ascertain the feasibility of the target invention

Learning the pertinent provisions of the Patent Laws

Filing of applications

> Learn how applications are presented and prosecuted

> Provisional application
>> Application Data Sheet with rough claims and drawings
>> Understand its significance
>>> Not the basis for prosecuting application
>>> Abandoned after one year
>>> It is only a notice of claim to an idea

> Non-Provisional application
>> Application Data Sheet
>>> Cover letter
>>> Fee calculation
>>> Oath
>>> Micro Filing Certification

>> Writing the specifications
>>> Specifications to support the claims
>>> Format for writing easily understood specification

# Introduction

Describing main, additional, and alternative embodiments
Specifications to tie in with claims
Abstract of Disclosure

## Drawings

Requirements for acceptable drawings
Different views of drawings
Rigid requirements for drawings

## Claims

Understand what independent and dependent claims are
Format for writing claims
Understand the use of antecedents
Resources to help in writing claims
Basis for prosecution of an application for a patent
Broad versus limiting claims
Amending claims

## Office Actions

Dealing with office actions
Responding to office actions
Arguing items in the office actions
Amending claims because of office actions
Overcoming emotions because of discouraging office actions

## PAIR (Patent Application Information Retrieval)

Requirements for participating in PAIR
Get application notarized
Get applicant's registration
Advantages of PAIR
Substantial reduction in filing fees
Instant retrieval of documents online

## Publication Requirements

## Allowability of claims and Patent maintenance

Fees dues
Keeping patent active
Infringement on patents

# Roadmap for Development of this Book

The Table of Contents developed by the United States Patent Offie itself helped in strictly following an organized path to writing this book, through a mass of provisions in the Patent Laws.

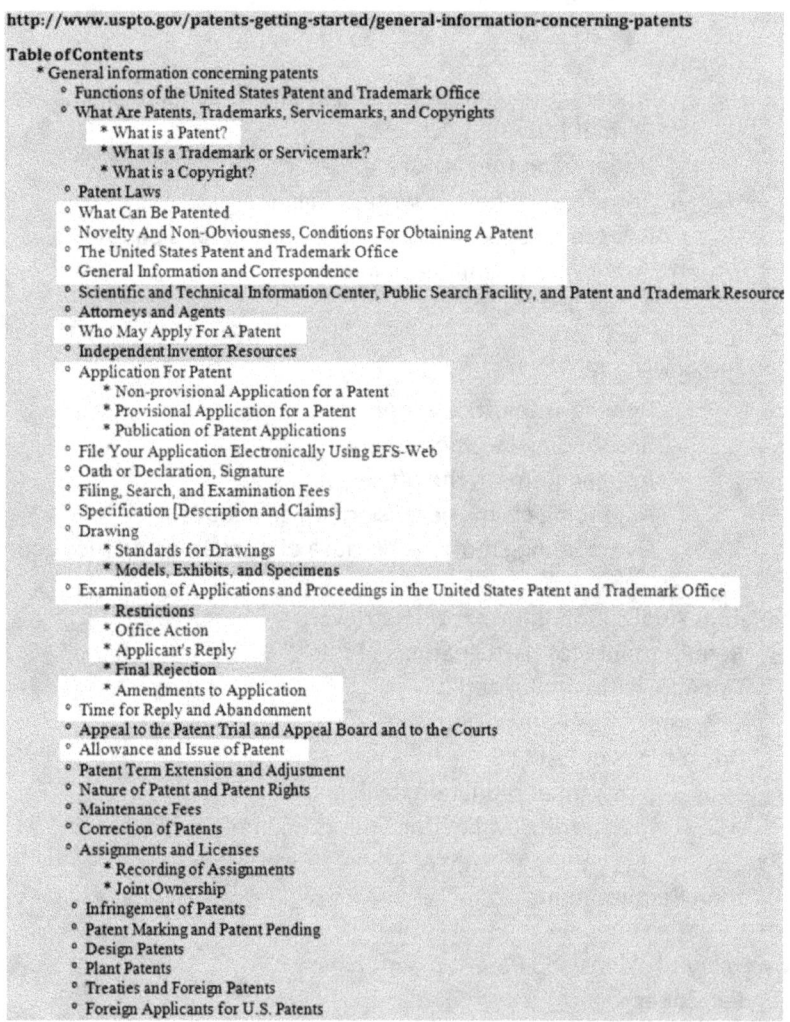

# Part 1 – The Goal is to initially present actual patent then show how it was secured

## Patent Certificate Cover Page 1

**Patent Certificate Cover Page 2**

## MAINTENANCE FEE NOTICE

*If the application for this patent was filed on or after December 12, 1980, maintenance fees are due three years and six months, seven years and six months, and eleven years and six months after the date of this grant, or within a grace period of six months thereafter upon payment of a surcharge as provided by law. The amount, number and timing of the maintenance fees required may be changed by law or regulation. Unless payment of the applicable maintenance fee is received in the United States Patent and Trademark Office on or before the date the fee is due or within a grace period of six months thereafter, the patent will expire as of the end of such grace period.*

## PATENT TERM NOTICE

*If the application for this patent was filed on or after June 8, 1995, the term of this patent begins on the date on which this patent issues and ends twenty years from the filing date of the application or, if the application contains a specific reference to an earlier filed application or applications under 35 U.S.C. 120, 121, 365(c), or 386(c), twenty years from the filing date of the earliest such application ("the twenty-year term"), subject to the payment of maintenance fees as provided by 35 U.S.C. 41(b), and any extension as provided by 35 U.S.C. 154(b) or 156 or any disclaimer under 35 U.S.C. 253.*

*If this application was filed prior to June 8, 1995, the term of*
*on the date on which this patent issues and ends on the later o*
*from the date of the grant of this patent or the twenty-year term*
*for patents resulting from applications filed on or after June 8,*
*the payment of maintenance fees as provided by 35 U.S.C.*
*extension as provided by 35 U.S.C. 156 or any disclaimer under*

Part 1 – The Goal is to initially present actual patent then show how it was secured

## Patent Certificate

## Patent Certificate Page 1

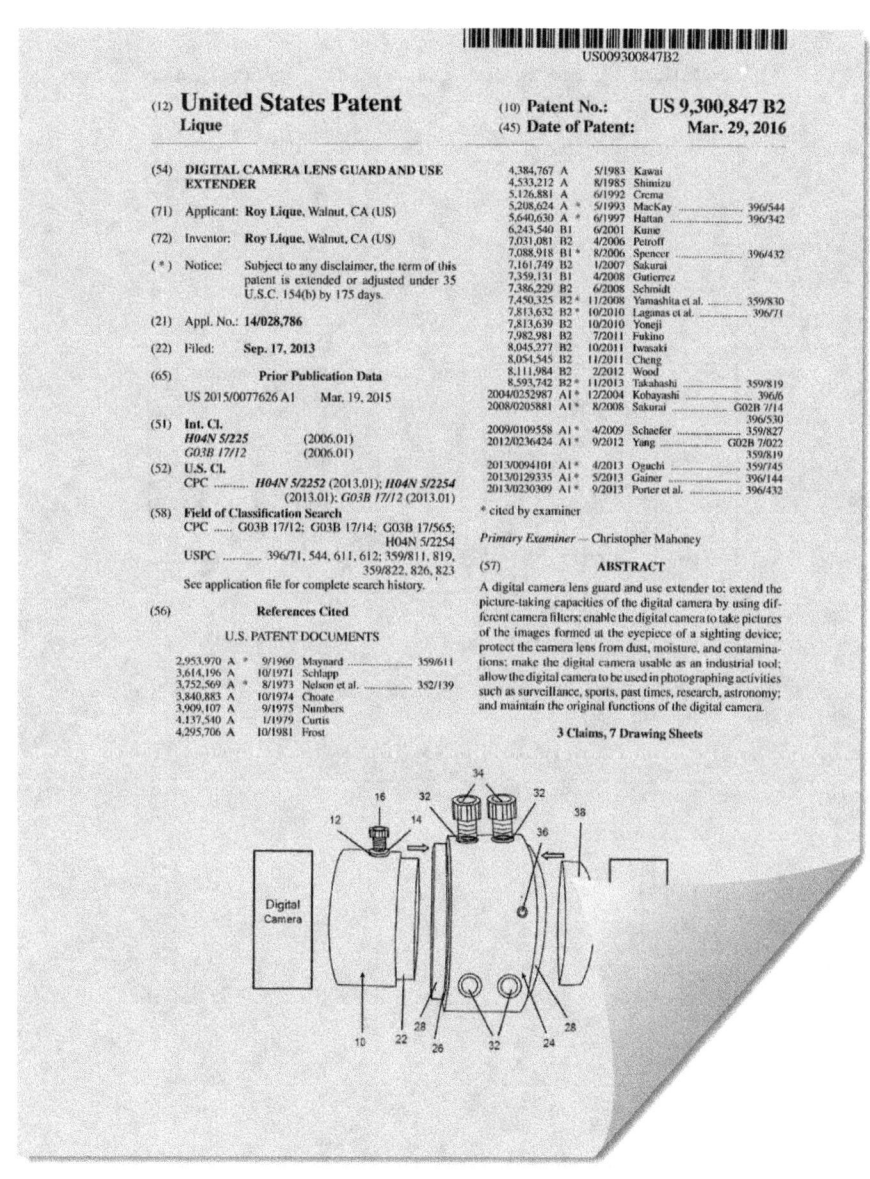

US009300847B2

(12) **United States Patent**
Lique

(10) Patent No.: **US 9,300,847 B2**
(45) Date of Patent: **Mar. 29, 2016**

(54) **DIGITAL CAMERA LENS GUARD AND USE EXTENDER**

(71) Applicant: **Roy Lique**, Walnut, CA (US)

(72) Inventor: **Roy Lique**, Walnut, CA (US)

(*) Notice: Subject to any disclaimer, the term of this patent is extended or adjusted under 35 U.S.C. 154(b) by 175 days.

(21) Appl. No.: **14/028,786**

(22) Filed: **Sep. 17, 2013**

(65) **Prior Publication Data**
US 2015/0077626 A1    Mar. 19, 2015

(51) **Int. Cl.**
*H04N 5/225*       (2006.01)
*G03B 17/12*       (2006.01)

(52) **U.S. Cl.**
CPC .......... *H04N 5/2252* (2013.01); *H04N 5/2254* (2013.01); *G03B 17/12* (2013.01)

(58) **Field of Classification Search**
CPC ...... G03B 17/12; G03B 17/14; G03B 17/565; H04N 5/2254
USPC .......... 396/71, 544, 611, 612; 359/811, 819, 359/822, 826, 823
See application file for complete search history.

(56) **References Cited**

U.S. PATENT DOCUMENTS

| | | | |
|---|---|---|---|
| 2,953,970 A * | 9/1960 | Maynard | 359/611 |
| 3,614,196 A | 10/1971 | Schlapp | |
| 3,752,569 A * | 8/1973 | Nelson et al. | 352/139 |
| 3,840,883 A | 10/1974 | Choate | |
| 3,909,107 A | 9/1975 | Numbers | |
| 4,137,540 A | 1/1979 | Curtis | |
| 4,295,706 A | 10/1981 | Frost | |
| 4,384,767 A | 5/1983 | Kawai | |
| 4,533,212 A | 8/1985 | Shimizu | |
| 5,126,881 A | 6/1992 | Crema | |
| 5,208,624 A * | 5/1993 | MacKay | 396/544 |
| 5,640,630 A * | 6/1997 | Hattan | 396/342 |
| 6,243,540 B1 | 6/2001 | Kume | |
| 7,031,081 B2 | 4/2006 | Petroff | |
| 7,088,918 B1 * | 8/2006 | Spencer | 396/432 |
| 7,161,749 B2 | 1/2007 | Sakurai | |
| 7,359,131 B1 | 4/2008 | Gutierrez | |
| 7,386,229 B2 | 6/2008 | Schmidt | |
| 7,450,325 B2 * | 11/2008 | Yamashita et al. | 359/830 |
| 7,813,632 B2 * | 10/2010 | Lagunas et al. | 396/71 |
| 7,813,639 B2 | 10/2010 | Yoneji | |
| 7,982,981 B2 | 7/2011 | Fukino | |
| 8,045,277 B2 | 10/2011 | Iwasaki | |
| 8,054,545 B2 | 11/2011 | Cheng | |
| 8,111,984 B2 | 2/2012 | Wood | |
| 8,593,742 B2 * | 11/2013 | Takahashi | 359/819 |
| 2004/0252987 A1 * | 12/2004 | Kobayashi | 396/6 |
| 2008/0205881 A1 * | 8/2008 | Sakurai | G02B 7/14 396/530 |
| 2009/0109558 A1 * | 4/2009 | Schaefer | 359/827 |
| 2012/0236424 A1 * | 9/2012 | Yang | G02B 7/022 359/819 |
| 2013/0094101 A1 * | 4/2013 | Oguchi | 359/745 |
| 2013/0129335 A1 * | 5/2013 | Gainer | 396/144 |
| 2013/0230309 A1 * | 9/2013 | Porter et al. | 396/432 |

* cited by examiner

*Primary Examiner* — Christopher Mahoney

(57)    **ABSTRACT**

A digital camera lens guard and use extender to: extend the picture-taking capacities of the digital camera by using different camera filters; enable the digital camera to take pictures of the images formed at the eyepiece of a sighting device; protect the camera lens from dust, moisture, and contaminations; make the digital camera usable as an industrial tool; allow the digital camera to be used in photographing activities such as surveillance, sports, past times, research, astronomy; and maintain the original functions of the digital camera.

**3 Claims, 7 Drawing Sheets**

Part 1 – The Goal is to initially present actual patent then show how it was secured

## Patent Certificate Page 2

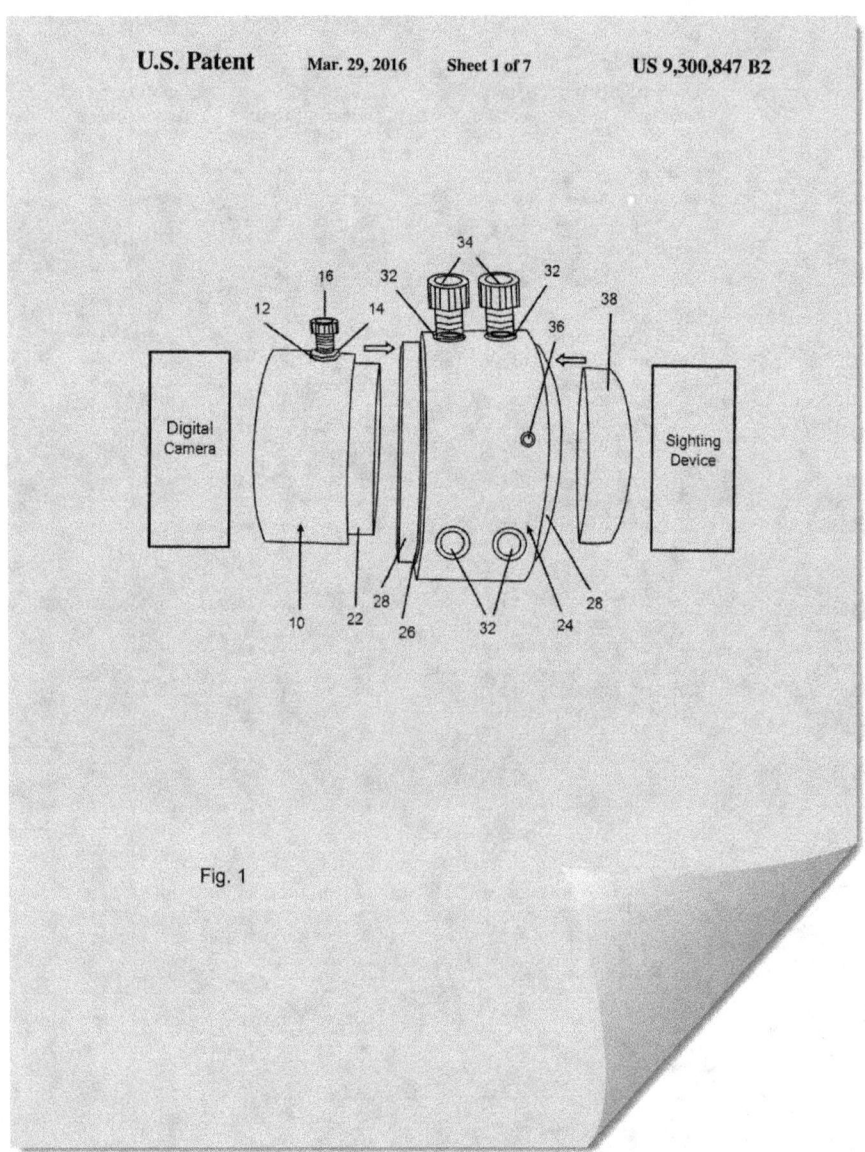

Fig. 1

**Patent Certificate Page 3**

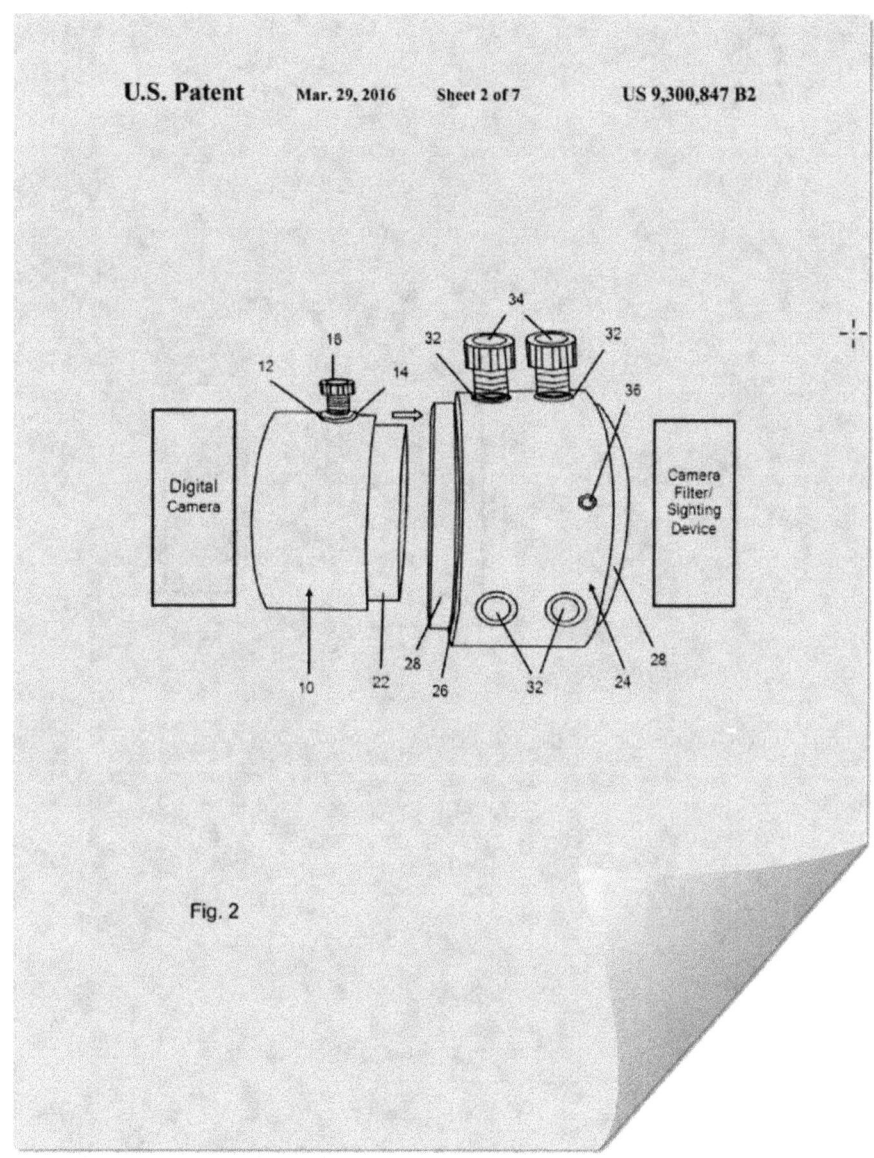

U.S. Patent        Mar. 29, 2016        Sheet 2 of 7        US 9,300,847 B2

Fig. 2

**Patent Certificate Page 4**

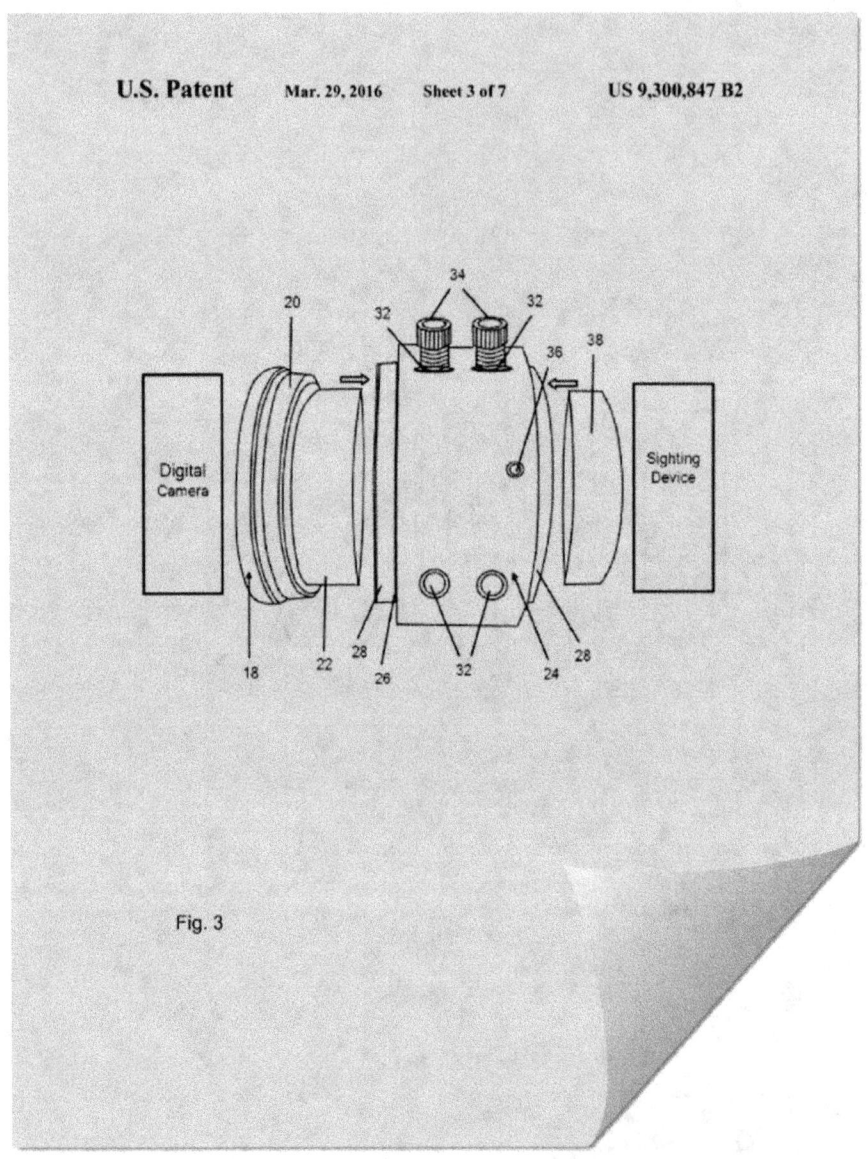

U.S. Patent   Mar. 29, 2016   Sheet 3 of 7   US 9,300,847 B2

Fig. 3

**Patent Certificate Page 5**

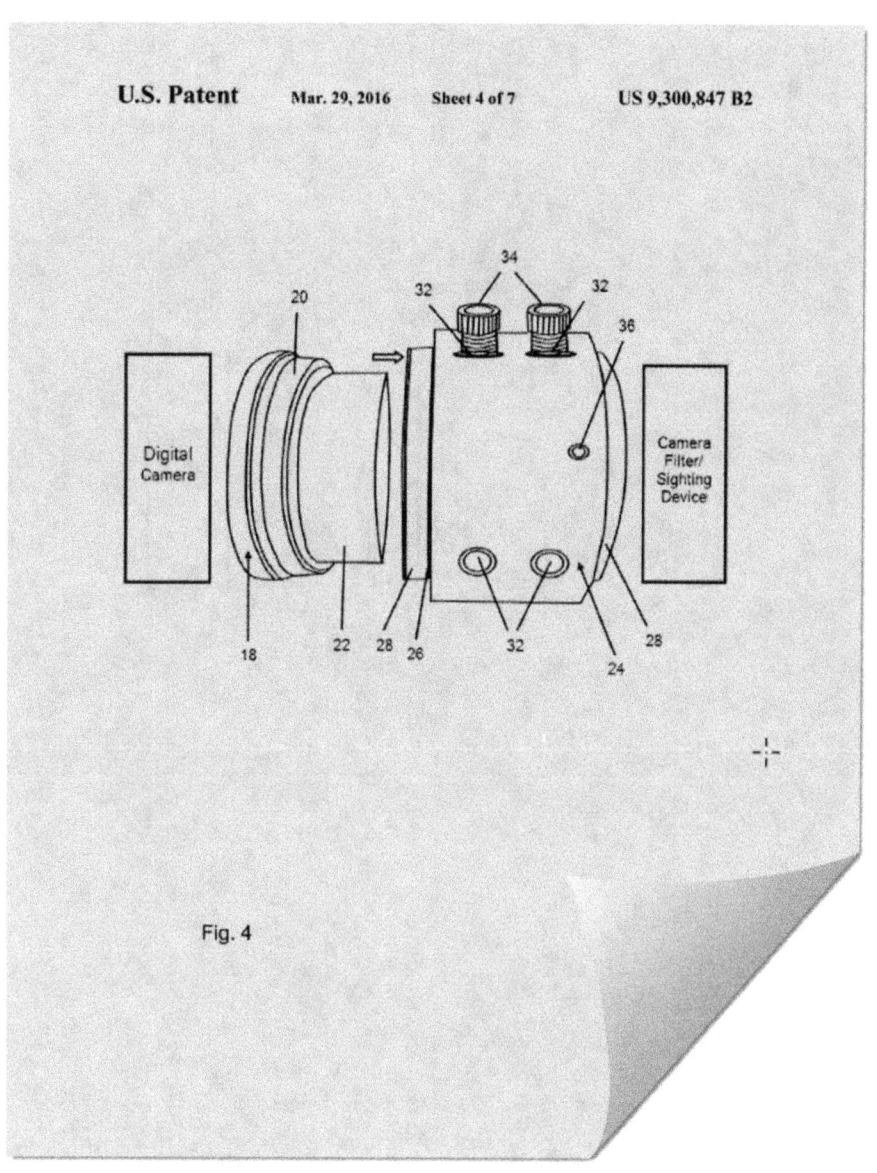

U.S. Patent      Mar. 29, 2016      Sheet 4 of 7      US 9,300,847 B2

Digital Camera

Camera Filter/ Sighting Device

Fig. 4

25

Part 1 – The Goal is to initially present actual patent then show how it was secured

**Patent Certificate Page 6**

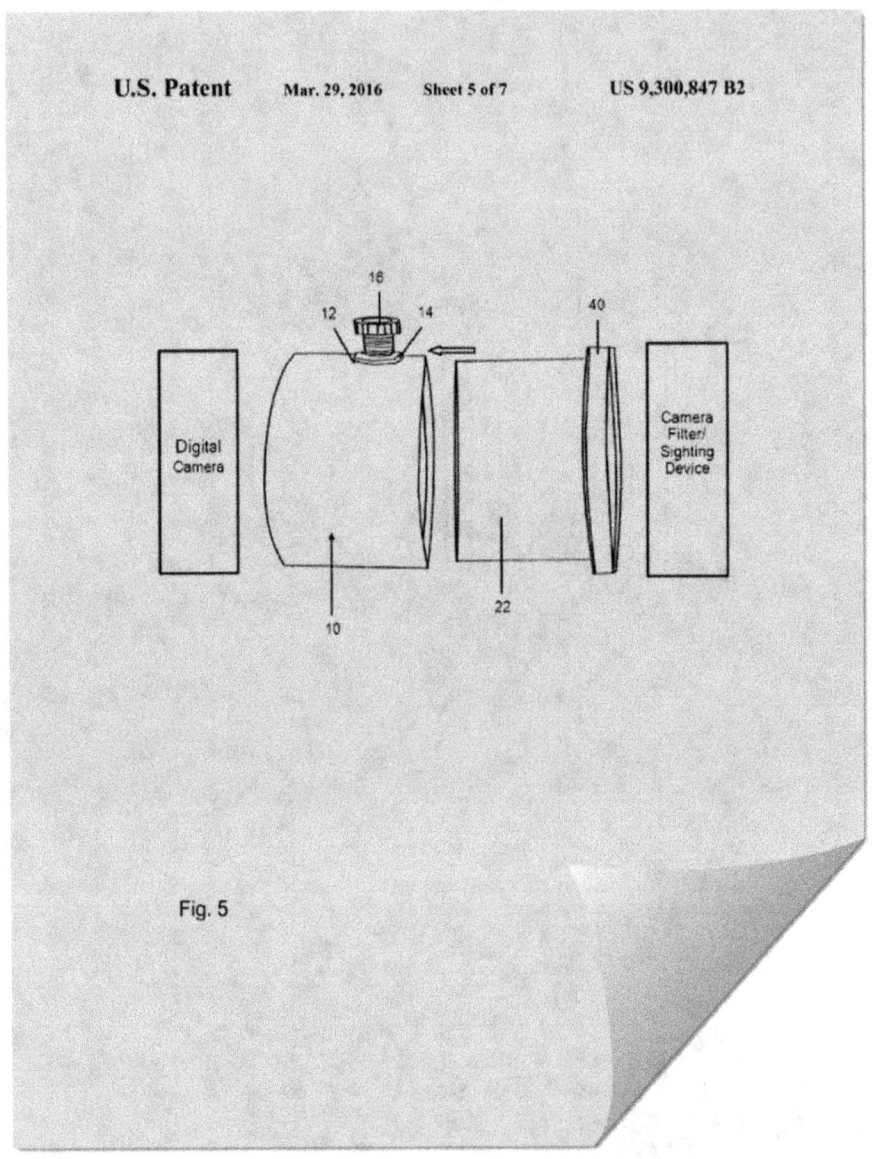

U.S. Patent    Mar. 29, 2016    Sheet 5 of 7    US 9,300,847 B2

Fig. 5

**Patent Certificate Page 7**

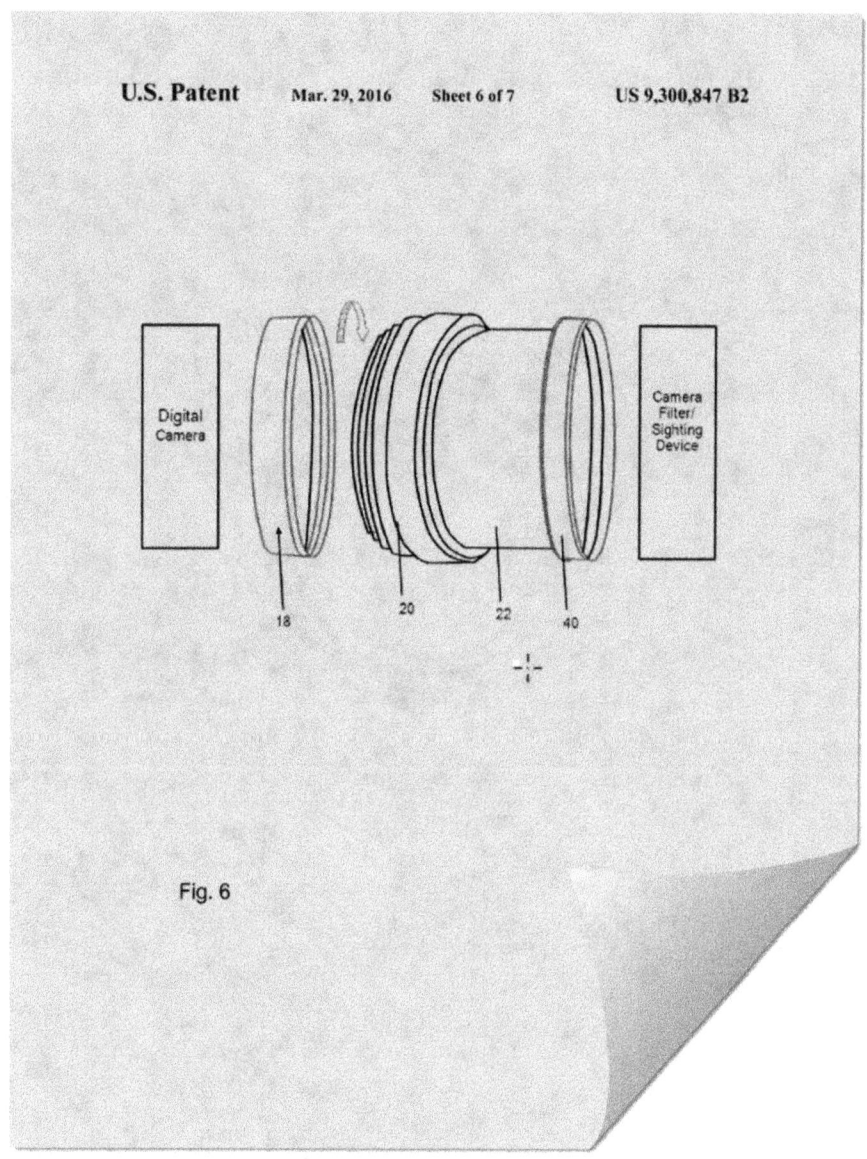

U.S. Patent     Mar. 29, 2016     Sheet 6 of 7     US 9,300,847 B2

Fig. 6

Part 1 – The Goal is to initially present actual patent then show how it was secured

## Patent Certificate Page 8

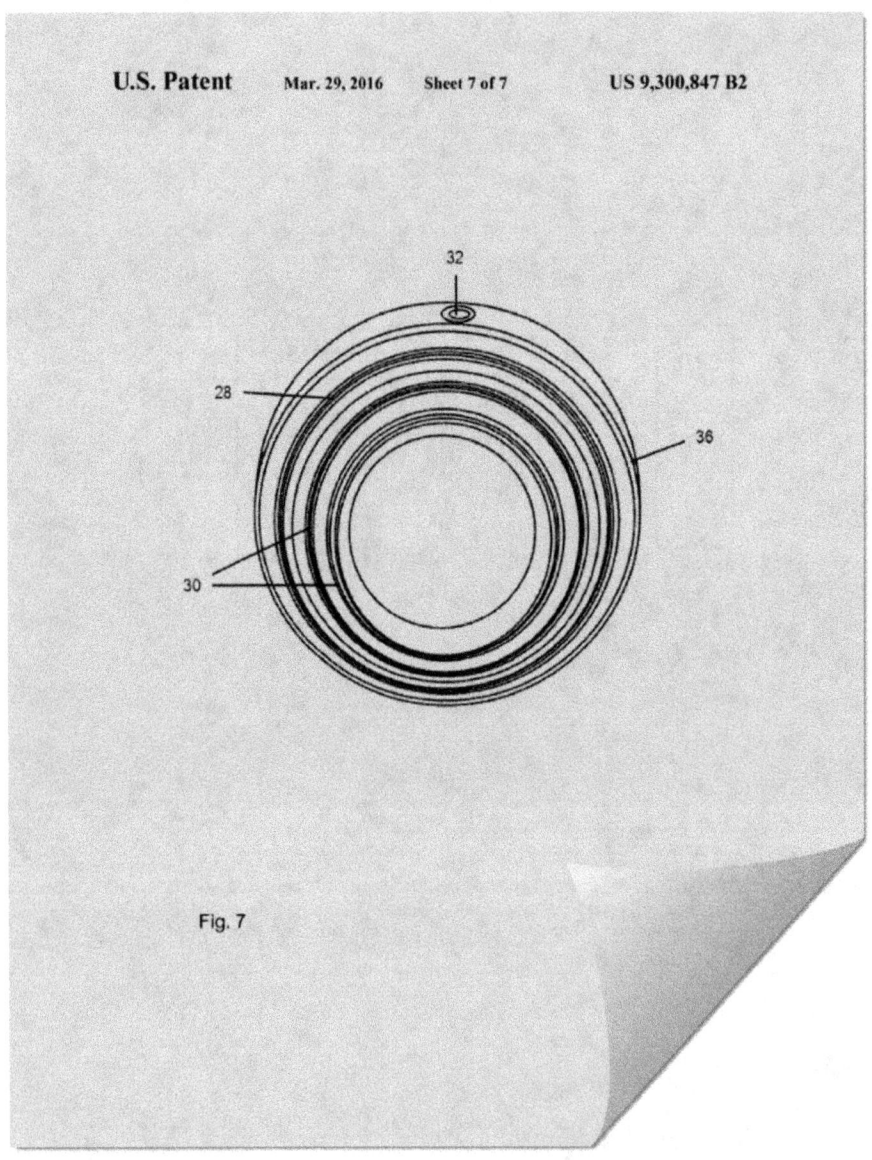

U.S. Patent     Mar. 29, 2016     Sheet 7 of 7     US 9,300,847 B2

Fig. 7

**Patent Certificate Page 9**

US 9,300,847 B2

**1**

### DIGITAL CAMERA LENS GUARD AND USE EXTENDER

#### CROSS REFERENCE TO RELATED APPLICATIONS

This application claims the benefit of provisional patent application Ser. No. 61/849,349, filed Jan. 25, 2013 by the present inventor.

#### BACKGROUND PRIOR ART

The following is a tabulation of some prior arts that presently appear barely relevant to the embodiments of the digital camera lens guard and use extender.

#### U.S. PATENTS

| Pat. No. | Title | Issue Date | Patentee |
|---|---|---|---|
| 8,111,984 | Matte box assembly | Feb. 7, 2012 | Wood; Dennis |
| 8,054,545 | Lens hood for a camera lens | Nov. 8, 2011 | Cheng; Ming-Chung |
| 7,813,039 | Camera cover | Oct. 12, 2010 | Yoneji; Osamu |
| 6,543,940 | Lens barrel assembly | Jun. 8, 2002 | Karter; Habloki |
| 5,126,881 | Lens hood for a photographic lens | Jun. 30, 1992 | Cretus; Rolf |
| 4,533,212 | Accessory holding device for optical instrument | Aug. 6, 1985 | Shiroim; Kenichi |
| 4,384,767 | Clamping device for camera accessory on head | May 24, 1983 | Kawai; Tokvo |
| 4,295,709 | Combined lens cap and sunshade for a camera | Oct. 20, 1981 | Frost; George H. |
| 4,137,540 | Camera matte box | Jan. 30, 1979 | Curtis; Jack |
| 3,909,207 | Hood for the lens of optical instruments with pivotally mounted lens cover | Sep. 30, 1975 | Nimmen; Jody L. |
| 3,840,880 | CAMERA LENS HOOD | Oct. 8, 1974 | Choate; J. Robert |
| 3,614,196 | COMBINED LENS HOOD AND FILTER SUPPORT | Oct. (9, 1971) | Schlage; Werner |

In the absence of significant relevance between prior arts and the embodiments of the digital camera lens guard and use extender, the use and benefits of the latter are discussed to create and present a new line of products principally dedicated to digital cameras.

The embodiments of the digital camera lens guard and use extender are in the field of cameras. More particularly, the embodiments extend and expand the capacities of a digital camera while also protecting the camera's lens. Capacities expansion is done with older as well as newer hardware, or a combination of both. Where applicable, expansion will also be done with software.

While the camera market is flooded with new camera models every year, it is also being depleted by obsolescence due to incompatibility with new software and hardware. It is also depleted by mere dislikes of older models and the accessories that come with them. Phone cameras also diminish the popularity of digital cameras. Some victims of obsolescence include multi-image lenses, fisheye lenses, macro and tele-photo lenses, square-shaped filters, over-sized and under-sized filters, fog and snow lenses, so on so forth.

Interchangeability of accessories between the digital camera and certain models of expensive cameras, is rare. The digital camera's lack of threads accounts for threaded accessories being almost exclusive monopolies of certain camera models. Some exciting photographs are taken using threaded accessories. Making the interchangeability problem more obvious is the fact that threads come in either metric or English.

**2**

Notwithstanding the additional sophisticated features coming with new cameras, there are still missed capacities that one would like to have in the digital camera. Examples are documenting an event happening too far from the viewer, or a past time too dangerous for the viewer to come close to the subject.

The additional new features that come with the later digital camera models include expensive electronics that need more protection. The entry of dust, moisture, and other contaminations into the lens area must be minimized in order to maintain the digital camera in an efficient working order. By engaging an embodiment of the digital camera lens guard and use extender to the digital camera and screwing in a camera filter, ample protection of the camera electronics is already provided.

The digital camera equipped with the embodiment of the digital camera lens guard and use extender, meets both the camera users' needs for additional camera capacities and lens and electronics protection without investing in expensive equipments.

#### SUMMARY

Functionally, the digital camera equipped with an embodiment of the digital camera lens guard and use extender compares with the more expensive models. The embodiment concentrates on enhancing the capacities of the digital camera and protecting its electronics.

With different size ada_____ _____ at the ends of the _____ ring of the embodiment, a s_____ _____ can be used. Obsolete and/o_____ because they can be used a_____ lenses from other camera mo_____ digital camera.

In conjunction with telesco_____ and other sighting devices, the _____ tool in industrial applications _____ become available for the digital _____ expanded capacities provided by _____

Aside from being used as indus_____ can now be used in other applica_____ crowd monitoring, event repor_____ research, and fast sports. Moreov_____ digital camera lens guard and us_____

US 9,300,847 B2

**3**

different camera brands and models. Consequently, investment in expensive equipments is postponed or skipped entirely for a while.

The digital camera is equipped with sensitive electronics. The chance of the digital camera being damaged due to the entry of dust, moisture, and other contaminations into the lens area is minimized with the use of the embodiment of the digital camera lens guard and use extender. Physical damage to the digital camera due to being dropped, bumped, and hit is also minimized. The digital camera remains portable despite the addition of the ring that comes with the camera mount assembly.

### DRAWINGS

#### Figures

FIG. 1 shows a lens guard with slip type camera mount assembly and a coupler ring, positioned between a camera and a sighting device.

FIG. 2 shows a lens guard with slip type camera mount assembly, positioned between a camera and a camera filter or a sighting device.

FIG. 3 shows a lens guard with screw type camera mount assembly and a coupler ring, positioned between a camera and a sighting device.

FIG. 4 shows a lens guard with screw type camera mount assembly, positioned between a camera and a camera filter or a sighting device.

FIG. 5 shows a slip type barrel track directly securing a camera filter or a sighting device, positioned between a camera and a sighting device.

FIG. 6 shows a screw type barrel track directly securing a camera filter or a sighting device, positioned between a camera and a sighting device.

FIG. 7 is an end view of a primary ring with a stack of outer and inner adapter rings.

### REFERENCE NUMERALS

| | |
|---|---|
| 10 — anchor ring | 12 — anchor hole |
| 14 — anchor nut | 16 — anchor screw |
| 18 — base adapter ring | 20 — track fastening ring |
| 22 — barrel track | 24 — primary ring |
| 26 — recess | 28 — outer adapter ring |
| 30 — inner adapter ring | 32 — retaining holes |
| 34 — retaining screws | 36 — insert holder |
| 38 — coupler ring | 40 — track end ring |

### GLOSSARY

Before the embodiments of the digital camera lens guard and use extender are described, some terms used here need to be defined.

The term "lens guard" refers to an embodiment of the digital camera lens guard and use extender.

The terms "slip type" and "screw type" refer to the action of connecting one component of the digital camera lens guard and use extender to another.

The terms "digital camera(s)" as used in these specifications, refers to a camera designed to be aimed to shoot pictures of optimal color, with ease and automatic adjustments of shutter speed, aperture, focus, and light sensitivity.

**4**

The terms "approximate" and "approximately" apply to numeric values and mean close to being exact.

The term "eyepiece" refers to the lens or lens group closest to the eye in an optical instrument.

The term "sighting device(s)" refers to a device used to assist in aligning or aiming weapons, surveying instruments, or other items by sight.

The term "wall" refers to the surface between the outside and inside diameters of a tube or cylinder.

The term "adapter ring(s)" refers to connectors for joining parts or devices having different sizes and designs, enabling them to be mated, fitted, and work together.

The term "camera filter(s)" refers to circular lens screen of plain or dyed gelatin or glass for controlling the rendering of color or for lessening the intensity of light and for protecting the camera lens.

### DETAILED DESCRIPTIONS

#### FIG. 1 and FIG. 2—First Embodiments

Construction of the embodiments of the digital camera lens guard and use extender described in these specifications is based on the camera barrel size range of 40 mm to 45 mm. All measurements, values, and dimensions are relative to this particular barrel size range. Other size ranges can be used provided their relationships of measurements, values, and dimensions are progressively and proportionally maintained.

Referring now to FIG. 1 and FIG. 2, there are shown embodiments of a digital camera lens guard and use extender having a camera mount assembly, a main housing assembly, and additionally in the case of FIG. 1, a coupler ring 38. The figures further show the position of each embodiment in relation to the digital camera represented by boxed "Digital Camera", and to the attachments represented by boxed "Sighting Device" or boxed "Camera Filter/Sighting Device". When used with the coupler ring 38, the embodiment offers only a single attachment and its position is shown in FIG. 1. With multiple allowable attachments attached one at a time, the position of the embodiment is shown in FIG. 2. The only difference between the embodiments shown in FIG. 1 and FIG. 2 is the presence or absence of the coupler ring 38.

Descriptions of the functions and construction details of the shown embodiments of the digital camera lens guard and use extender follow.

Camera Mount Assembly.

Still referring to FIG. 1 and FIG. 2, an anchor ring 10 of the camera mount assembly is shown as the base that secures the embodiment of the digital camera lens guard and use extender to the digital camera. It is also the basis for most of the sizes, values, measurements, and dimensions of the other components of the embodiment.

Slip Type Camera Mou...

In more details, still ref...
type camera mount assemb...
an anchor hole 12, an anch...
a barrel track 22.

The action of slipping the ...
ring 10 after the anchor ring...
camera, is what "slip type" cam...
The action, simultaneous with...
barrel 22 to the primary ring 24...
the digital camera lens guard an...
camera.

Anchor Ring 10.

The cylindrical anchor ring 1...
(0.512") in length, with approx...

30

**Patent Certificate Page 11**

US 9,300,847 B2

5

inside diameter, and with approximately 1.5 mm (0.058") wall is cut from 6061 grade aluminum. The wall provides sufficient thickness for attaching the anchor ring 10 upright to the digital camera concentrically with the camera barrel. The anchor hole 12 facilitates the attachment of a pair of sufficiently large anchor screw 16 and anchor nut 14, to the anchor ring 10. The anchor screw 16 and the anchor nut 14 are used to secure the barrel track 22.

Barrel Track 22.

The cylindrical barrel track 22 is also from 6061 grade aluminum, cut to an approximate length of 25.4 mm (1.0"). Approximately one half of its entire length is inserted into the anchor ring 10 to support a secure connection. The other approximate half serves as a flange that inserts into the primary ring 24. The depth of insertion of the flange into the primary ring 24 is adjustable, making it useful in minimizing the formation of circles around pictures taken by the digital camera.

Main Housing Assembly.

In more details still referring to FIG. 1 and FIG. 2, the main housing assembly comprises a cylindrical primary ring 24 with appurtenances. It is in the main housing assembly that the major components of the embodiment of the digital camera lens guard and use extender come together and form a secure connection. Photographing activities take place in the main housing assembly.

Primary Ring 24.

The primary ring 24 accepts and secures a camera filter or a sighting device through the embedded inner adapter ring 30 (not shown) or outer adapter ring 28 at its ends. Additionally in FIG. 1, the primary ring 24 accepts and secures a sighting device through a coupler ring 38.

The primary ring 24 is also cut from 6061 grade aluminum to approximately 25.4 mm (1.0") long. Sufficient space for embedding the commonly used size ranges of adapter rings is provided by its approximate outside diameter of 63.5 mm (2.5") and approximate wall of 6.4 mm (0.252"). Its approximate inside diameter of 44.5 mm (1.752") provides the additional track for the camera barrel to extend and retract without obstruction.

Owing to the identical ends of the primary ring 24 with different size adapter rings, the directions at which the ends point are reversible, offering more opportunities for attaching different camera filters or sighting devices, one at a time.

Adapter Rings.

Size-matched and gender-matched camera filter or sighting device is accepted at either end of the primary ring 24. The embedded adapter rings at the ends of the primary ring 24 immediately provide more opportunities to use different camera filters and sighting devices, one at a time. Since circles around images taken by the digital camera are sometimes caused by the addition of loose adapter rings and camera filters, care is observed that only enough of them are used as needed.

The inside and outside diameters of a target adapter ring are modified as necessary to sufficient sizes so that the adapter ring can be embedded at either end of the primary ring 24. Optionally, its male threads are stripped off. Two types of adapter rings are optionally embedded at either end of the primary ring 24, namely, inner adapter ring 30 and outer adapter ring 28. Their construction details are described as follows:

Inner Adapter Ring 30.

To attach the inner adapter ring 30 at either end of the primary ring 24, an approximately 6.4 mm (0.252") deep recess 26 with sufficient circumference to accept the target inner adapter ring 30, is carved. The inner adapter ring 30 is

6

attached resting at the bottom of the recess 26 with the female threads oriented outwards, using industry grade adhesive.

Outer Adapter Ring 28.

To attach the outer adapter ring 28 at either end of the primary ring 24, an optional approximately 3.2 mm (0.125") deep recess 26 with sufficient circumference to accept the target outer adapter ring 28, is carved. The outer adapter ring 28 is attached resting at the bottom of the recess 26 with the female threads oriented outwards, using industry grade adhesive.

An alternative way to embed the outer adapter ring 28 into the primary ring 24 is to cleanly cut off the recessed portion of the primary ring 24, referring to previous paragraph. Using industry grade adhesive, the outer adapter ring 28 is inserted and attached flushed with the end of the cut off portion. The cut off portion is attached back to the primary ring 24 making sure the female threads are oriented outwards.

The outer adapter ring 28 can also be attached directly to either end of the primary ring 24 without carving the recess 26. It only needs to be stripped off of its male threads and attached to the primary ring 24 with the female threads oriented outwards, using industry grade adhesive.

Stack of Adapter Rings.

Building a stack of adapter rings as shown in FIG. 7, follows the steps described for embedding the inner adapter ring 30 and the outer adapter ring 28. The inner adapter rings 30 progressively get smaller as the stack grows, each inner adapter ring 30 resting approximately 6.4 mm (0.252") deeper from the preceding larger one.

Retaining Holes 32 and Retaining Screws 34.

In order to provide ample handling surface, retaining holes 32 of sufficient size are drilled approximately 6.4 mm (0.252") from either end of the primary ring 24. Because at least one retaining screw 34 is used to secure the barrel track 22 and another one to secure the coupler ring 38, the locations of the retaining holes 32 are sufficiently far apart to allow the retaining screws 34 to turn freely.

Insert Holes 36.

The optional insert holes 36 are of sufficient size to accept inserts for expansion and improvements. They are drilled onto the primary ring 24 at desired locations that do not interfere with the outer adapter rings 28, inner adapter rings 30, and retaining screws 34.

Coupler Ring 38.

In more details referring to FIG. 1, the coupler ring 38 provides the connection between the main housing assembly and the sighting device. The sighting device is secured to the main housing assembly by properly inserting its eyepiece into the coupler ring 38 and the coupler ring 38 into the primary ring 24.

The coupler ring 38 is of sufficient size and length, preferably between 25.4 mm (1.0") and 50.8 mm (2.0"), and has an outside diameter closely matching the inside diameter of the primary ring 24. The inside diameter of the coupler ring 38 varies depending on the size of the eyepiece of the target sighting device. If necessary, an adapter ring is embedded to alter its inside diameter in order to closely match the size of the eyepiece of the sighting device. The coupler ring 38 is also cut from 6061 grade aluminum.

The construction details of the embodiments of the digital camera lens guard and use extender referring to FIG. 1 and FIG. 2 are that the embodiments can be made from any other sufficiently rigid and strong material such as high strength plastic and the like. Further, the construction details of the embodiments of the digital camera lens guard and use extender can be made of different materials from different sources, brands, and styles.

**Patent Certificate Page 12**

US 9,300,847 B2

**7**

Operation

FIG. 1 and FIG. 2

Referring to FIG. 1 and FIG. 2, either end of the barrel track 22 is inserted into the anchor ring 10. Anchor screw 16 is tightened to secure the barrel track 22.

The open end of the barrel track 22 is inserted into either end of the primary ring 24. It is inserted opposite the end where the camera filter or sighting device is attached. The size and type of camera filter or sighting device determine which end of the primary ring 24 needs to accept and secure the barrel track 22.

Insertion depth of the barrel track 22 into the primary ring 24 is fixed for the distance the camera barrel has to extend. One or more retaining screws 34 are tightened to secure the barrel track 22.

In more details referring to FIG. 1, the eyepiece of a sighting device is inserted into either end of the coupler ring 38. If the outside circumference of the eyepiece is smaller than the inside circumference of the coupler ring 38, a fitting (not shown) is used to make the insertion snugly secure. The open end of the coupler ring 38 is inserted as far as it can go or until it is stopped by the barrel track 22, into the primary ring 24.

Alternatively, the use of the coupler ring 38 can be done away with as shown in FIG. 2. Size-matched and gender-matched camera filter or sighting device is screwed directly to either the outer adapter ring 28 or the inner adapter ring 30 (not shown). If necessary, loose adapter rings are added to find a match between the embedded adapter rings and the target camera filter or sighting device.

In more details referring to FIG. 1 and FIG. 2, as shown, the embodiments of the digital camera lens guard and use extender provide for quick and easy installation of the camera filter or the sighting device to the main housing assembly. They also facilitate their quick and easy mounting and dismounting to and from the digital camera. More size-matched and gender-matched camera filters and sighting devices unavailable with the digital camera before, become available now with the use of the outer adapter rings 28 and the inner adapter rings 30.

FIG. 3 and FIG. 4

Additional Embodiments

Referring now to FIG. 3 and FIG. 4, there are shown embodiments of the digital camera lens guard and use extender having a camera mount assembly, a main housing assembly, and additionally in the case of FIG. 3, a coupler ring 38. The figures further show the position of each embodiment in relation to the digital camera represented by boxed "Digital Camera", and to the attachments represented by boxed "Sighting Device" or boxed "Camera Filter/Sighting Device". When used with the coupler ring 38, the embodiment offers only a single attachment and its position is shown in FIG. 3. With multiple allowable attachments attached one at a time, the position of the embodiment is shown in FIG. 4. The only difference between the embodiments shown in FIG. 3 and FIG. 4 is the presence or absence of the coupler ring 38.

Descriptions of the functions and construction details of the shown embodiments of the digital camera lens guard and use extender follow.

Camera Mount Assembly.

Still referring to FIG. 3 and FIG. 4, a base adapter ring 18 of the camera mount assembly secures the embodiment of the digital camera lens guard and use extender to the digital

**8**

camera. It is the basis for most of the sizes, values, measurements, and dimensions of the other components of the embodiments.

Screw Type Camera Mount Assembly.

In more details still referring to FIG. 3 and FIG. 4, the screw type camera mount assembly comprises the modified base adapter ring 18 and the configured barrel track 22.

The action of screwing the track mounting ring 20 to the modified base adapter ring 18 after the base adapter ring 18 is mounted on the digital camera, is what "screw type" camera mount assembly refers to. The action, simultaneous with slipping the other end of the barrel 22 to the primary ring 24, secures the embodiment of the digital camera lens guard and use extender to the digital camera.

Base Adapter Ring 18.

Initially, the inside diameter of the base adapter ring 18 is approximately 40 mm (1.575") to 45 mm (1.772") and its outside diameter is approximately 50 mm (1.969"). The sizes are suitable for modification by machining, to circumferentially enclose the camera barrel. With its male threads stripped off, the base adapter ring 18 is attached to the digital camera upright concentrically with the camera barrel and with the female threads oriented outwards, using industry grade adhesive. To allow the camera barrel to extend and retract without obstruction, a space is maintained between the camera barrel and the base adapter ring 18.

Barrel Track 22.

The barrel track 22 is also cut from 6061 grade aluminum to an approximate length of 25.4 mm (1.0"). Using industry grade adhesive, one end is fitted with the modified track mounting ring 20 which is size-matched and gender-matched with the base adapter ring 18. The track mounting ring 20 is made flush with the end of the barrel track 22. The other end serves as a flange that inserts into the primary ring 24. The depth of insertion of the flange into the primary ring 24 is adjustable, making it useful in minimizing the formation of circles around pictures taken by the digital camera.

Main Housing Assembly.

The main housing assembly is identical to that described in the embodiments shown in FIG. 1 and FIG. 2.

Coupler Ring 38.

The coupler ring 38 is identical to that described in the embodiments shown in FIG. 1 and FIG. 2.

The construction details of the embodiments of the digital camera lens guard and use extender as shown in FIG. 3 and FIG. 4 are that the embodiments may be made of metal or of any other sufficiently rigid and strong material such as high-strength plastic and the like. Further, the various components of the embodiments of the digital camera lens guard and use extender can be made o... sources, brands, and style...

Op...

FIG. 3 a...

Referring to FIG. 3 and FIG... is screwed into the base adapter...

The open end of the barrel tr... end of the primary ring 24. It i... where the camera filter or sighting... and type of camera filter or sight... end of the primary ring 24 need... barrel track 22.

**Patent Certificate Page 13**

US 9,300,847 B2

**9**

Insertion depth of the barrel track 22 into the primary ring 24 is fixed for the distance the camera barrel has to extend. One or more retaining screws 34 are tightened to secure the barrel track 22.

In more detail referring to FIG. 3, the eyepiece of a sighting device is inserted into either end of the coupler ring 38. If the outside circumference of the eyepiece is smaller than the inside circumference of the coupler ring 38, a fitting (not shown) is used to make the insertion snugly secure. The open end of the coupler ring 38 is inserted as far as it can go or until it is stopped by the barrel track 22, into the primary ring 24.

Alternatively, the use of the coupler ring 38 can be done away with as shown in FIG. 4. Size-matched and gender-matched camera filter or sighting device is screwed directly to either the outer adapter ring 28 or the inner adapter ring 30 (not shown). If necessary, loose adapter rings are added to find a match between the embedded adapter rings and the target camera filter or sighting device.

In more detail referring to FIG. 3 and FIG. 4, as shown, the embodiments of the digital camera lens guard and use extender provide for quick and easy installation of the camera filter or the sighting device to the main housing assembly. They also facilitate their quick and easy mounting and dismounting to and from the digital camera. More size-matched and gender-matched camera filters and sighting devices unusable with the digital camera before become available now with the use of the outer adapter rings 28 and the inner adapter rings 30.

### FIG. 5 and FIG. 6

#### Alternative Embodiments

Referring to FIG. 5 and FIG. 6, there are shown simplified embodiments of the digital camera lens guard and use extender. The figures further show the position of each embodiment in relation to the digital camera represented by boxed "Digital Camera". With multiple allowable attachments represented by boxed "Camera Filter/Sighting Device" attached one at a time, the relative positions of the embodiments are also shown in both figures.

In FIG. 5 the end of the barrel track 22 opposite the end that slips into the anchor ring 10 is fitted with a track end ring 40 with the female threads oriented outwards. In FIG. 6, fitting is done at the end opposite that which screws into the base adapter ring 18. In both instances, fitting is done using industry grade adhesive and ensuring that the track end ring 40 is flushed with the end of the barrel track 22. The camera filter or sighting device is screwed directly into the track end ring 40.

The construction details of the embodiments of the digital camera lens guard and use extender as shown in FIG. 5 and FIG. 6 are that the embodiments may be made of metal or of any other sufficiently rigid and strong material such as high-strength plastic and the like. Further, the various components of the embodiments of the digital camera lens guard and use extender can be made of different materials from different sources, brands, and styles.

Referring to FIG. 5 and FIG. 6, as shown, the embodiments of the digital camera lens guard and use extender provide for simple and direct use of camera filters and sighting devices by completely bypassing the main housing assembly.

#### Advantages

Broadly, from the description above, a number of advantages of most embodiments of the digital camera lens guard and use extender become evident:

**10**

(a) The embodiments of the digital camera lens guard and use extender have the advantage of possibly being one of the few dedicated to digital cameras.

(b) Due to its simple design, future changes on the embodiments of the digital camera lens guard and use extender will be easily implemented.

(c) The features added by sighting devices make the digital camera more adaptable to various photo opportunities.

(d) With the added features of a sighting device, the digital camera can be used as an industrial tool.

(e) Mounting and dismounting of an embodiment to and from the digital camera takes only few turns of the track mounting ring or the anchor screw.

(f) The digital camera remains portable despite the addition of a base ring.

(g) Embodiments of the digital camera lens guard and use extender of different sizes can be manufactured for different digital camera types, models, and sighting devices, provided appropriate matching adapter rings are used.

(h) The digital cameras are now able to take pictures previously possible only with the more expensive cameras.

(i) Camera protection extends the life of the digital camera and is accomplished with just a few turns of a camera filter.

(j) Camera users will benefit from innovations and improvements from two different industrial classifications, namely, digital cameras and sighting devices.

(k) With the different size adapter rings fitted at the ends of the primary ring, immediately a large number of camera filters becomes available.

(l) Tripods are used less frequently because of the nature of digital cameras.

#### CONCLUSION, RAMIFICATIONS, AND SCOPE

Accordingly, a digital camera user will see that a digital camera enabled by the embodiments of the digital camera lens guard and use extender is well adapted to various photo taking sessions. As the embodiments are easy to mount and dismount to and from the digital camera, more photographic events can be recorded.

In terms of functionalities, the market availability of camera filters and sighting devices makes the digital camera comparable with the more expensive types and models. With the option to choose which end of the main housing assembly to attach camera filters and sighting devices, the possibilities become more numerous.

More specifically, the following are few examples of the use of the digital camera enabled by the embodiments of the digital camera lens guard and use extender:

surveillance from a di... when getting close embodying the observer,

observing a phenomenon...

research such as observi...

crowd observation such as...

monitoring such as vehicle...

past time such as bird-wat...

safaris such as observing wi...

emergency reporting such as...

fire fighting such as reporting...

progress report such as of a n...

fast sports such as tennis mat...

traffic surveillance such as mo...

weather observation such as a...

crowd safety such as life guar...

law enforcement such as unru...

**Patent Certificate Page 14**

US 9,300,847 B2

**11**

While the foregoing written descriptions of the embodiments of the digital camera lens guard and use extender enables one of ordinary skill to make and use what is considered presently to be the best mode thereof, those of ordinary skills will understand and appreciate the existence of variations, combinations, and equivalents of the specific embodiment, method, and examples herein. The digital camera lens guard and use extender should therefore not be limited by the above described embodiments, methods, and examples, but by all embodiments and methods within the scope and spirit of the digital camera lens guard and use extender as claimed.

I claim:

1. A digital camera lens guard and use extender for adding deviant picture-taking capacities to a digital camera, and protecting said camera's lens, barrel, and electronics, comprising:

    a. a camera mount assembly having a circular base and a cylindrical ring of sufficient size and length and a plurality of locking means, for allowing unobstructed movement of said digital camera's barrel,

    b. a main housing assembly having a primary ring of sufficient size and length with a plurality of retaining

**12**

screws and embed locations for adapter rings on each end, for union with camera filters and sighting devices.

2. The digital camera lens guard and use extender of claim 1 wherein said embed location is a circular recess carved approximately one-eighth to one-fourth inch at each said end of said primary ring of said main housing assembly.

3. The digital camera lens guard and use extender of claim 1 wherein said primary ring of said main housing assembly being inserted with a coupler ring, for securing an eyepiece of a sighting device,

    whereby the digital camera lens guard and use extender adds picture-taking capacities to said digital camera beside that of point-and-shoot method; provides protection for said digital camera's lens, barrel, and electronics; facilitates usage of camera filters, adapter rings, and sighting devices; enables said digital camera to take pictures of images formed at said eyepiece of a sighting device; and allows said digital camera to accept new picture-taking capacities that are introduced in the form of adapter rings.

\* \* \* \* \*

# Part 2 – Quick Course on Patent Laws

## What is a Patent

A patent for an invention is the grant of a property right to the inventor, issued by the United States Patent and Trademark Office. Generally, the term of a new patent is 20 years from the date on which the application for the patent was filed in the United States or, in special cases, from the date an earlier related application was filed, subject to the payment of maintenance fees. U.S. patent grants are effective only within the United States, U.S. territories, and U.S. possessions. Under certain circumstances, patent term extensions or adjustments may be available.

The right conferred by the patent grant is, in the language of the statute and of the grant itself, "the right to exclude others from making, using, offering for sale, or selling" the invention in the United States or "importing" the invention into the United States. What is granted is not the right to make, use, offer for sale, sell or import, but the right to exclude others from making, using, offering for sale, selling or importing the invention. Once a patent is issued, the patentee must enforce the patent without aid of the USPTO.

There are three types of patents:

1)Utility patents may be granted to anyone who invents or discovers any new and useful process, machine, article of manufacture, or composition of matter, or any new and useful improvement thereof;

2) Design patents may be granted to anyone who invents a new, original, and ornamental design for an article of manufacture; and

3) Plant patents may be granted to anyone who invents or discovers and asexually reproduces any distinct and new variety of plant.

The patent law specifies the general field of subject matter that can be patented and the conditions under which a patent may be obtained.

In the language of the statute, any person who "invents or discovers any new and useful process, machine, manufacture, or composition of matter, or any new and useful improvement thereof, may obtain a patent," subject to the conditions and requirements of the law. The word "process" is defined by law as a process, act, or method, and primarily includes industrial or technical processes. The term "machine" used in the statute needs no explanation. The term "manufacture" refers to articles that are made, and includes all manufactured articles. The term "composition of matter" relates to chemical compositions and may include mixtures of ingredients as well as new chemical compounds. These classes of subject matter taken together include practically everything that is made by man and the processes for making the products.

The Atomic Energy Act of 1954 excludes the patenting of inventions useful solely in the utilization of special nuclear material or atomic energy in an atomic weapon. See 42 U.S.C. 2181(a).

The patent law specifies that the subject matter must be "useful." The term "useful" in this connection refers to the condition that the subject matter has a useful purpose and also includes operativeness, that is, a machine which will not operate to perform the intended purpose would not be called useful, and therefore would not be granted a patent.

Interpretations of the statute by the courts have defined the limits of the field of subject matter that can be patented, thus it has been held that the laws of nature, physical phenomena, and abstract ideas are not patentable subject matter.

A patent cannot be obtained upon a mere idea or suggestion. The patent is granted upon the new machine, manufacture, etc., as has been said, and not upon the idea or suggestion of the new machine. A complete description of the actual machine or other subject matter for which a patent is sought is required.

## Novelty And Non-Obviousness, Conditions For Obtaining A Patent

In order for an invention to be patentable it must be new as defined in the patent law, which provides that an invention cannot be patented if:

"(1) the claimed invention was patented, described in a printed publication, or in public use, on sale, or otherwise available to the public before the effective filing date of the claimed invention" or

"(2) the claimed invention was described in a patent issued [by the U.S.] or in an application for patent published or deemed published [by the U.S.], in which the patent or application, as the case may be, names another inventor and was effectively filed before the effective filing date of the claimed invention."

There are certain limited patent law exceptions to patent prohibitions (1) and (2) above. Notably, an exception may apply to a "disclosure made 1 year or less before the effective filing date of the claimed invention," but only if "the disclosure was made by the inventor or joint inventor or by another who obtained the subject matter disclosed… from the inventor or a joint inventor."

In patent prohibition (1), the term "otherwise available to the public" refers to other types of disclosures of the claimed invention such as, for example, an oral presentation at a scientific meeting, a demonstration at a trade show, a lecture or speech, a statement made on a radio talk show, a YouTube™ video, or a website or other on-line material.

Effective filing date of the claimed invention: This term appears in patent prohibitions (1) and (2). For a U.S. nonprovisional patent application that is the first application containing the claimed subject matter, the term "effective filing date of the claimed invention" means the actual filing date of the U.S. nonprovisional patent application. For a U.S. nonprovisional application that claims the benefit of a corresponding prior-filed U.S. provisional application, "effective filing date of the claimed invention" can be the filing date of the prior-filed provisional application provided the provisional application sufficiently describes the claimed invention. Similarly, for a U.S. nonprovisional application that is a continuation or division of a prior-filed U.S. nonprovisional application, "effective filing date of the claimed invention" can be the filing date of the prior filed nonprovisional application that sufficiently describes the claimed invention. Finally, "effective filing date of the claimed invention" may be the filing date of a prior-filed foreign patent application to which foreign priority is claimed provided the foreign patent application sufficiently describes the claimed invention.

Even if the subject matter sought to be patented is not exactly shown by the prior art, and involves one or more differences over the most nearly similar thing already known, a patent may still be refused if the differences would be obvious. The subject matter sought to be patented must be sufficiently different from what has been used or described before that it may be said to be non-obvious to a person having ordinary skill in the area of technology related to the invention. For example, the substitution of one color for another, or changes in size, are ordinarily not patentable.

## The United States Patent and Trademark Office

Congress established the United States Patent and Trademark Office to issue patents on behalf of the government. The Patent Office as a distinct bureau dates from the year 1802 when a separate official in the Department of State, who became known as "Superintendent of Patents," was placed in charge of patents. The revision of the patent laws enacted in 1836 reorganized the Patent Office and designated the official in charge as Commissioner of Patents. The Patent Office remained in the Department of State until 1849 when it was transferred to the Department of Interior. In 1925 it was transferred to the Department of Commerce where it is today. The name of the Patent Office was changed to the Patent and Trademark Office in 1975 and changed to the United States Patent and Trademark Office in 2000.

The USPTO administers the patent laws as they relate to the granting of patents for inventions, and performs other duties relating to patents. Applications for patents are examined to determine if the applicants are entitled to patents under the law and patents are granted when applicants are so entitled. The USPTO publishes issued patents and most patent

applications 18 months from the earliest effective application filing date, and makes various other publications concerning patents. The USPTO also records assignments of patents, maintains a search room for the use of the public to examine issued patents and records, and supplies copies of records and other papers, and the like. Similar functions are performed with respect to the registration of trademarks. The USPTO has no jurisdiction over questions of infringement and the enforcement of patents.

The head of the Office is the Under Secretary of Commerce for Intellectual Property and Director of the United States Patent and Trademark Office (Director). The Director's staff includes the Deputy Under Secretary of Commerce and Deputy Director of the USPTO, the Commissioner for Patents, the Commissioner for Trademarks, and other officials. As head of the Office, the Director superintends or performs all duties respecting the granting and issuing of patents and the registration of trademarks; exercises general supervision over the entire work of the USPTO; prescribes the rules, subject to the approval of the Secretary of Commerce, for the conduct of proceedings in the USPTO, and for recognition of attorneys and agents; decides various questions brought before the Office by petition as prescribed by the rules; and performs other duties necessary and required for the administration of the United States Patent and Trademark Office.

The work of examining applications for patents is divided among a number of examining technology centers (TCs), each TC having jurisdiction over certain assigned fields of technology. Each TC is headed by group directors and staffed by examiners and support staff. The examiners review applications for patents and determine whether patents can be granted. An appeal can be taken to the Patent Trial and Appeal Board from their decisions refusing to grant a patent, and a review by the Director of the USPTO may be had on other matters by petition. In addition to the examining TCs, other offices perform various services, such as receiving and distributing mail, receiving new applications, handling sales of printed copies of patents, making copies of records, inspecting drawings, and recording assignments.

At present, the USPTO has over 11,000 employees, of whom about three quarters are examiners and others with technical and legal training. Patent applications are received at the rate of over 500,000 per year.

Effective November 15, 2011, any regular nonprovisional utility application filed by mail or hand-delivery will require payment of an additional $400 fee called the "non-electronic filing fee," which is reduced by 50 percent (to $200) for applicants that qualify for small entity status under 37 CFR 1.27(a). The 75 percent micro entity discount does not apply to the non-electronic filing fee and consequently the non-electronic filing fee is also $200 for applicants that qualify for micro entity status under 37 CFR 1.29(a) or (d). This fee is required by Section 10(h) of the Leahy-Smith America Invents Act, Public Law 112-29 (Sept. 16, 2011; 125 Stat. 284). The only way to avoid having to pay the additional $400 non-electronic filing fee is to file the regular nonprovisional utility patent application via EFS-Web.Design, plant, and provisional applications are not subject to the additional non-electronic filing fee and may continue to be filed by mail or hand-delivery without additional charge. See the information available at www.uspto.gov/patents/process/file/efs/index/jsp. Any questions regarding filing applications via EFS-Web should be directed to the Electronic Business Center at 866-217-9197.

## General Information and Correspondence

All business with the United States Patent and Trademark Office should be transacted in writing. Regular nonprovisional utility applications must be filed via EFS-Web in order to avoid the additional $400 non-electronic filing fee.

Other patent correspondence, including design, plant, and provisional application filings, as well as correspondence filed in a nonprovisional application after the application filing date (known as "follow-on" correspondence), can still be filed by mail or hand-delivery without incurring the $400 non-electronic filing fee.

Such other correspondence relating to patent matters should be addressed to

**COMMISSIONER FOR PATENTS

P.O. Box 1450

Alexandria, VA 22313-1450**

when sent by mail via the United States Postal Service. If a mail stop is appropriate, the mail stop should also be used.

Mail addressed to different mail stops should be mailed separately to ensure proper routing. For example, after final correspondence should be mailed to

**Mail Stop AF

Commissioner for Patents

P.O. Box 1450

Alexandria, VA 22313-1450**

and assignments should be mailed to

**Mail Stop Assignment Recordation Services

Director of the U.S. Patent and Trademark Office

P.O. Box 1450

Alexandria, VA 22313-1450**

Correspondents should be sure to include their full return addresses, including zip codes. The principal location of the USPTO is 600 Dulany Street, Alexandria, Virginia. The personal presence of applicants at the USPTO is unnecessary.

You do not have to be a Registered eFiler to file a patent application via EFS-Web. However, unless you are a Registered eFiler, you must not attempt to file follow-on correspondence via EFS-Web, because Unregistered eFilers are not permitted to file follow-on correspondence via

EFS-Web. Follow-on correspondence filed by anyone other than an EFS-Web Registered eFiler must be sent by mail or hand-delivered to the address specified in the paragraph above.

Applicants and attorneys are required to conduct their business with decorum and courtesy. Papers presented in violation of this requirement will be returned.

Separate letters (but not necessarily in separate envelopes) should be written for each distinct subject of inquiry, such as assignments, payments, orders for printed copies of patents, orders for copies of records, and requests for other services. None of these inquiries should be included with letters responding to Office actions in applications.

When a letter concerns a patent application, the correspondent must include the application number (consisting of the series code and the serial number, e.g., 12/123,456) or the serial number and filing date assigned to that application by the Office, or the international application number of the international application number of the international application. When a letter concerns a patent (other than for purposes of payment of a maintenance fee), it should include the name of the patentee, the title of the invention, the patent number, and the date of issue.

An order for a copy of an assignment should identify the reel and frame number where the assignment or document is recorded; otherwise, an additional charge is made for the time consumed in making the search for the assignment.

Applications for patents, which are not published or issued as patents, are not generally open to the public, and no information concerning them is released except on written authority of the applicant, his or her assignee, or his or her attorney, or when necessary to the conduct of the business of the USPTO. Patent application publications and patents and related records, including records of any decisions, the records of assignments other than those relating to assignments of unpublished patent applications, patent applications that are relied upon for priority in a patent application publication or patent, books, and other records and papers in the Office are open to the public. They may be inspected in the USPTO Search Room, or copies may be ordered.

The Office cannot respond to inquiries concerning the novelty and patentability of an invention prior to the filing of an application; give advice as to possible infringement of a patent; advise of the propriety of filing an application; respond to inquiries as to whether, or to whom, any alleged invention has been patented; act as an expounder of the patent law or as counselor for individuals, except in deciding questions arising before it in regularly filed cases. Information of a general nature may be furnished either directly or by supplying or calling attention to an appropriate publication.

# Who May Apply For A Patent

According to the law, the inventor, or a person to whom the inventor has assigned or is under an obligation to assign the invention, may apply for a patent, with certain exceptions. If the inventor is deceased, the application may be made by legal representatives, that is, the administrator or executor of the estate. If the inventor is legally incapacitated, the application for patent may be made by a legal representative (e.g., guardian). If an inventor refuses to apply for a patent or cannot be found, a joint inventor may apply on behalf of the non-signing inventor.

If two or more persons make an invention jointly, they apply for a patent as joint inventors. A person who makes only a financial contribution is not a joint inventor and cannot be joined in the application as an inventor. It is possible to correct an innocent mistake in erroneously omitting an inventor or in erroneously naming a person as an inventor.

Officers and employees of the United States Patent and Trademark Office are prohibited by law from applying for a patent or acquiring, directly or indirectly, except by inheritance or bequest, any patent or any right or interest in any patent.

## Application For Patent

### Non-provisional Application for a Patent

A nonprovisional application for a patent is made to the Director of the United States Patent and Trademark Office and includes:

(1) A written document which comprises a specification (description and claims);

(2) Drawings (when necessary);

(3) An oath or declaration; and

(4) Filing, search, and examination fees. Fees for filing, searching, examining, issuing, appealing, and maintaining patent applications and patents are reduced by 50 percent for any small entity that qualifies for reduced fees under 37 CFR 1.27(a), and are reduced by 75 percent for any micro entity that files a certification that the requirements under 37 CFR 1.29(a) or (d) are met.

Small Entity Status: Applicant must determine that small entity status under 37 CFR 1.27(a) is appropriate before making an assertion of entitlement to small entity status and paying a fee at the 50 percent small entity discount. Fees change each October. Note that by filing electronically via EFS-Web, the filing fee for an applicant qualifying for small entity status is further reduced.

Micro Entity Status: Applicant must determine that micro entity status under 37 CFR 1.29(a) or (d) is appropriate before filing the required certification of micro entity status and paying a fee at the 75 percent micro entity discount. The patent forms Web page is indexed under the section titled Forms, Patents on the USPTO website at www.uspto.gov. There are two micro entity certification forms – namely form PTO/SB/15A for certifying micro entity status on the "gross income basis" under 37 CFR 1.29(a), and form PTO/SB/15B for certifying micro entity status on the "institution of higher education basis" under 37 CFR 1.29(d). Effective November 15, 2011, any regular nonprovisional utility application filed by mail or hand-delivery will require payment of an additional $400 fee called the "non-electronic filing fee," which is reduced by 50 percent (to $200) for applicants that qualify for small entity status under 37 CFR 1.27(a) or micro entity status under 37 CFR 1.29(a) or (d). The only way to avoid having to pay the additional $400 non-electronic filing fee is by filing the regular nonprovisional utility application via EFS-Web.

Other patent correspondence, including design, plant, and provisional application filings, as well as correspondence filed in a nonprovisional application after the application filing date (known as "follow-on" correspondence), can still be filed by mail or hand-delivery without incurring the $400 non-electronic filing fee. You do not have to be a Registered eFiler to file a patent application via EFS-Web. However, unless you are a Registered eFiler, you must not attempt to file follow-on correspondence via EFS-Web, because Unregistered eFilers are not permitted to file follow-on correspondence via EFS-Web. Follow-on correspondence filed by anyone other than an EFS-Web Registered eFiler must be sent by mail or be hand-delivered. (See the "General Information and Correspondence" section of this brochure.) In the event you receive from the USPTO a "Notice of Incomplete Application" in response to your EFS-Web filing stating that an application number has been assigned but no filing date has been granted, you must become a Registered eFiler and file your reply to the "Notice of Incomplete Application" via EFS-Web in order to avoid the $400 non-electronic filing fee. To become a Registered eFiler and have the ability to file follow-on correspondence, please consult the information at www.uspto.gov/patents/process/file/efs/guidance/register.jsp, or call the Electronic Business Center at 866-217-9197.

The specification (description and claims) can be created using a word processing program such as Microsoft® Word or Corel® WordPerfect. The document containing the specification can normally be converted into PDF format by the word processing program itself so that it can be included as an attachment when filing the application via EFS-Web. Other application documents, such as drawings and a hand-signed declaration, may have to be scanned as a PDF file for filing via EFS-Web. See the information available at www.uspto.gov/patents/process/file/efs/index.jsp. Any questions regarding filing applications via EFS-Web should be directed to the Electronic Business Center at 866-217-9197.

All application documents must be in the English language or a translation into the English language will be required along with the required fee set forth in 37 CFR 1.17(i).

Each document (which should be filed via EFS-Web in PDF format) must have a top margin of at least 2 cm (3/4 inch), a left side margin of at least 2.5 cm (1 inch), a right side margin of at least 2 cm (3/4 inch) and a bottom margin of at least 2 cm (3/4 inch) with no holes made in the submitted papers. It is also required that the spacing on all papers be 1.5 or double-spaced and the application papers must be numbered consecutively (centrally located above or below the text) starting with page one.

The specification must have text written in a nonscript font (e.g., Arial, Times Roman, or Courier, preferably a font size of 12pt) lettering style having capital letters that should be at least 0.3175 cm (0.125 inch) high, but may be no smaller than 0.21 cm (0.08 inch) high (e.g., a font size of 6). The specification must have only a single column of text.

The specification must conclude with a claim or claims particularly pointing out and distinctly claiming the subject matter that the applicant regards as the invention. The portion of the application in which the applicant sets forth the claim or claims is an important part of the application, as it is the claims that define the scope of the protection afforded by the patent. The claims must commence on a separate sheet.

More than one claim may be presented provided they differ from each other. Claims may be presented in independent form (e.g. the claim stands by itself) or in dependent form, referring back to and further limiting another claim or claims in the same application. Any dependent claim that refers back to more than one other claim is considered a "multiple dependent claim."

The application for patent is not forwarded for examination until all required parts, complying with the rules related thereto, are received. If any application is filed without all the required parts for obtaining a filing date (incomplete or defective), the applicant will be notified of the deficiencies and given a time period to complete the application filing (a surcharge may be required)—at which time a filing date as of the date of such a completed submission will be obtained by the applicant. If the omission is not corrected within a specified time period, the application will be returned or otherwise disposed of; the filing fee if submitted will be refunded less a handling fee as set forth in the fee schedule.

The filing fee and declaration or oath need not be submitted with the parts requiring a filing date. It is, however, desirable that all parts of the complete application be deposited in the Office together; otherwise, each part must be signed and a letter must accompany each part, accurately and clearly connecting it with the other parts of the application. If an application that has been accorded a filing date does not include the filing fee or the oath or declaration, applicant will be notified and given a time period to pay the filing fee, file an oath or declaration and pay a surcharge.

All applications received in the USPTO are numbered in sequential order, and the applicant will be informed of the application number and filing date by a filing receipt.

The filing date of an application for patent is the date on which a specification (including at least one claim) and any drawings necessary to understand the subject matter sought to be patented are received in the USPTO; or the date on which the last part completing the application is received in the case of a previously incomplete or defective application.

## Provisional Application for a Patent

Since June 8, 1995, the USPTO has offered inventors the option of filing a provisional application for patent, which was designed to provide a lower-cost first patent filing in the United States and to give U.S. applicants parity with foreign applicants. Claims and oath or declaration are NOT required for a provisional application. A provisional application provides the means to establish an early effective filing date in a patent application and permits the term "Patent Pending" to be applied in connection with the invention. Provisional applications may not be filed for design inventions.

The filing date of a provisional application is the date on which a written description of the invention, and drawings if necessary, are received in the USPTO. To be complete, a provisional application must also include the filing fee, and a cover sheet specifying that the application is a provisional application for patent. The applicant would then have up to 12 months to file a nonprovisional application for patent as described above. The claimed subject matter in the later filed nonprovisional application is entitled to the benefit of the filing date of the provisional application if it has support in the provisional application.

If a provisional application is not filed in English, and a nonprovisional application is filed claiming benefit to the provisional application, a translation of the provisional application will be required. See title 37, Code of Federal Regulations, Section 1.78(a)(5).

Provisional applications are NOT examined on their merits. A provisional application will become abandoned by the operation of law 12 months from its filing date. The 12-month pendency for a provisional application is not counted toward the 20-year term of a patent granted on a subsequently filed nonprovisional application that claims benefit of the filing date of the provisional application.

A surcharge is required for filing the basic filing fee or the cover sheet on a date later than the filing of the provisional application. Unlike nonprovisional utility applications, design, plant, and provisional applications can still be filed by mail or hand-delivery without having to pay the additional $400 non-electronic filing fee. Design and provisional applications can also be filed via EFS-Web. Plant applications, however, are not permitted to be filed via EFS-Web.

## Publication of Patent Applications

Publication of patent applications is required by the American Inventors Protection Act of 1999 for most plant and utility patent applications filed on or after November 29, 2000. On filing of a plant or utility application on or after November 29, 2000, an applicant may request that the application not be published, but only if the invention has not been and will not be the subject of an application filed in a foreign country that requires publication 18 months after filing (or earlier claimed priority date) or under the Patent Cooperation Treaty.

Publication occurs after the expiration of an 18-month period following the earliest effective filing date or priority date claimed by an application. Following publication, the application for patent is no longer held in confidence by the Office and any member of the public may request access to the entire file history of the application.

As a result of publication, an applicant may assert provisional rights. These rights provide a patentee with the opportunity to obtain a reasonable royalty from a third party that infringes a published application claim provided actual notice is given to the third party by applicant, and patent issues from the application with a substantially identical claim. Thus, damages for pre-patent grant infringement by another are now available.

# File Your Application Electronically Using EFS-Web

Effective November 15, 2011, any regular nonprovisional utility application filed by mail or hand-delivery will require payment of an additional $400 fee called the "non-electronic filing fee," which is reduced by 50 percent (to $200) for applicants that qualify for small entity status under 37 CFR 1.27(a) or micro entity status under 37 CFR 1.29(a) or (d). The only way to avoid having to pay the additional $400 non-electronic filing fee is by filing your nonprovisional utility application via EFS-Web. A small entity applicant who files electronically not only avoids the additional non-electronic filing ($200 for small entity and micro entity applicants); the small entity applicant who files electronically also receives a bigger discount on the regular filing fee. Any questions regarding filing applications via EFS-Web should be directed to the Electronic Business Center at 866-217-9197.

http://www.uspto.gov/patents-getting-started/general-information-con

Other patent correspondence, including design, plant, and provisional application filings, as well as correspondence filed in a nonprovisional application after the application filing date (known as "follow-on" correspondence), can still be filed by mail or hand-delivery without incurring the $400 non-electronic filing fee. You do not have to be a Registered eFiler to file a patent application via EFS-Web. However, unless you are a Registered eFiler, you must not attempt to file follow-on correspondence via EFS-Web, because Unregistered eFilers are not permitted to file follow-on correspondence via EFS-Web. Follow-on correspondence filed by anyone other than an EFS-Web Registered eFiler must be sent by mail or be hand-delivered. (See the "General Information and Correspondence" section of this brochure.) In the event you receive from the USPTO a "Notice of Incomplete Application" in response to your EFS-Web filing stating that an application number has been assigned but no filing date has been granted, you must become a Registered eFiler and file your reply to the "Notice of Incomplete Application" via EFS-Web in order to avoid the $400 non-electronic filing fee. To become a Registered eFiler and have the ability to file follow-on correspondence, please consult the information at www.uspto.gov/patents/process/file/efs/guidance/register.jsp, or call the Electronic Business Center at 866-217-9197.

EFS-Web allows customers to electronically file patent application documents securely via the Internet via a Web page. EFS-Web is a system for submitting new applications and documents related to previously-filed patent applications. Customers prepare documents in Portable Document Format (PDF), attach the documents, validate that the PDF documents will be compatible with USPTO internal automated information systems, submit the documents, and pay fees with real-time payment processing. Some forms are available as fillable EFS-Web forms. When these fillable EFS-Web forms are used, the data entered into the forms is automatically loaded into USPTO information systems.

EFS-Web can be used to submit:

(A) New utility patent applications and fees

(B) New design patent applications and fees

(C) Provisional patent applications and fees

(D) Requests to enter the national stage under 35 U.S.C. 371 and fees

(E) Most follow-on documents and fees for a previously filed patent application

Further information on EFS-Web is available at www.uspto.gov/patents/process/file/efs/guidance.

See the "Legal Framework" document on that Web page for a list of correspondence that may not be filed via EFS-Web and answers to frequently asked questions.

## Oath or Declaration, Signature

An oath or declaration is a formal statement that must be made by the inventor in a non-provisional application. Each inventor must sign an oath or declaration that includes certain statements required by law and the USPTO rules, including the statement that he or she believes himself or herself to be the original inventor or an original joint inventor of a claimed invention in the application and the statement that the application was made or authorized to be made by him or her. See 35 U.S.C 115 and 37 CFR 1.63. An oath must be sworn to by the inventor before a notary public. A declaration may be submitted in lieu of an oath. A declaration does not need to be notarized. Oaths or declarations are required for design, plant, utility, and reissue applications. In addition to the required statements, the oath or declaration must set forth the legal name of the inventor, and, if not provided in an application data sheet, the inventor's mailing address and residence. In lieu of an oath or declaration, a substitute statement may be signed by the applicant with respect to an inventor who is deceased, legally incapacitated, cannot be found or reached after diligent effort, or has refused to execute the oath or declaration. When filing a continuing application, a copy of the oath or declaration filed in the earlier application may be used provided that it complies with the rules in effect for the continuing application (i.e., the rules that apply to applications filed on or after September 16, 2012).

Forms for declarations are available by calling the USPTO General Information Services at 800-786-9199 or 571-272-1000 or by accessing USPTO website at www.uspto.gov, indexed under the section titled "Forms, Patents." Most of the forms on the USPTO website are electronically fillable and can be included in the application filed via EFS-Web without having to print the form out in order to scan it for inclusion as a PDF attachment to the application.

## Filing, Search, and Examination Fees

A patent application is subject to the payment of a basic fee and additional fees that include a search fee, an examination fee, and issue fee. Consult the USPTO website atwww.uspto.gov for the current fees. Total claims that exceed 20, and independent claims that exceed three are considered "excess claims" for which additional fees are due. For example, if applicant filed a total of 25 claims, including four independent claims, applicant would be required to pay excess claims fees for five total claims exceeding 20, and one independent claim exceeding three. If the same applicant later filed an amendment increasing the total number of claims to 29, and the number of independent claims to six, applicant would be required to pay more excess claims fees for the four additional total claims and the two additional independent claims.

In calculating fees, a claim is singularly dependent if it incorporates by reference a single preceding claim that may be an independent or dependent claim. A multiple dependent claim or any claim depending therefrom shall be considered as separate dependent claims in accordance with the number of claims to which reference is made. In addition, if the application contains multiple dependent claims, an additional fee is required for each multiple dependent claim.

If the owner of the invention is a small entity, (an independent inventor, a small business concern or a nonprofit organization), most fees are reduced by half if small entity status is claimed. If small entity status is desired and appropriate, applicants should pay the small entity filing fee. Applicants claiming small entity status should make an investigation as to whether small entity status is appropriate before claiming such status.

Most of the fees are subject to change in October of each year.

## Specification [Description and Claims]

The following order of arrangement should be observed in framing the application:

(a) Application transmittal form

(b) Fee transmittal form

(c) Application Data Sheet

(d) Specification

(e) Drawings

(f) Executed Oath or declaration

The specification should have the following sections, in order:

(1) Title of the Invention

(2) Cross Reference to related applications (if any). (Related applications may be listed on an application data sheet, either instead of or together with being listed in the specification.)

(3) Statement of federally sponsored research or development (if any)

(4) The names of the parties to a joint research agreement if the claimed invention was made as a result of activities within the scope of a joint research agreement

(5) Reference to a "Sequence Listing," a table, or a computer program listing appendix submitted on a compact disc and an incorporation by reference of the material on the compact disc. The total number of compact disc including duplicates and the files on each compact disc shall be specified.

(6) Background of the Invention

(7) Brief Summary of the Invention

(8) Brief description of the several views of the drawing (if any)

(9) Detailed Description of the Invention

(10) A claim or claims

(11) Abstract of the disclosure

(12) Sequence listing (if any)

The specification must include a written description of the invention and of the manner and process of making and using it, and is required to be in such full, clear, concise, and exact terms as to enable any person skilled in the technological area to which the invention pertains, or with which it is most nearly connected, to make and use the same.

The specification must set forth the precise invention for which a patent is solicited, in such manner as to distinguish it from other inventions and from what is old. It must describe completely a specific embodiment of the process, machine, manufacture, composition of matter, or improvement invented, and must explain the mode of operation or principle whenever applicable. The best mode contemplated by the inventor for carrying out the invention must be set forth.

In the case of an improvement, the specification must particularly point out the part or parts of the process, machine, manufacture, or composition of matter to which the improvement relates, and the description should be confined to the specific improvement and to such parts as necessarily cooperate with it or as may be necessary to a complete understanding or description of it.

The title of the invention, which should be as short and specific as possible (no more than 500 characters), should appear as a heading on the first page of the specification if it does not otherwise appear at the beginning of the application. A brief abstract of the technical disclosure in the specification, including that which is new in the art to which the invention pertains, must be set forth on a separate page preferably following the claims. The abstract should be in the form of a single paragraph of 150 words or less, under the heading "Abstract of the Disclosure."

A brief summary of the invention indicating its nature and substance, which may include a statement of the object of the invention, should precede the detailed description. The summary should be commensurate with the invention as claimed, and any object recited should be that of the invention as claimed.

When there are drawings, there shall be a brief description of the several views of the drawings, and the detailed description of the invention shall refer to the different views by specifying the numbers of the figures, and to the different parts by use of reference numerals.

The specification must conclude with a claim or claims particularly pointing out and distinctly claiming the subject matter that the applicant regards as the invention. The portion of the application in which the applicant sets forth the claim or claims is an important part of the application, as it is the claims that define the scope of the protection afforded by the patent and which questions of infringement are judged by the courts.

More than one claim may be presented, provided they differ substantially from each other and are not unduly multiplied. One or more claims may be presented in dependent form, referring back to and further limiting another claim or claims in the same application. Any dependent claim that refers back to more than one other claim is considered a "multiple dependent claim."

Multiple dependent claims shall refer to such other claims in the alternative only. A multiple dependent claim shall not serve as a basis for any other multiple dependent claims. Claims in dependent form shall be construed to include all of the limitations of the claim incorporated by reference into the dependent claim. A multiple dependent claim shall be construed to incorporate all the limitations of each of the particular claims in relation to which it is being considered.

The claim or claims must conform to the invention as set forth in the remainder of the specification and the terms and phrases used in the claims must find clear support or antecedent basis in the description so that the meaning of the terms in the claims may be ascertainable by reference to the description.

## Specifications

The specification should have the following sections, in order:

(1) Title of the Invention

(2) Cross Reference to related applications (if any). (Related applications may be listed on an application data sheet, either instead of or together with being listed in the specification.)

(3) Statement of federally sponsored research or development (if any)

(4) The names of the parties to a joint research agreement if the claimed invention was made as a result of activities within the scope of a joint research agreement

(5) Reference to a "Sequence Listing," a table, or a computer program listing appendix submitted on a compact disc and an incorporation by reference of the material on the compact disc. The total number of compact disc including duplicates and the files on each compact disc shall be specified.

(6) Background of the Invention

(7) Brief Summary of the Invention

(8) Brief description of the several views of the drawing (if any)

(9) Detailed Description of the Invention

(10) A claim or claims

(11) Abstract of the disclosure

(12) Sequence listing (if any)

The specification must include a written description of the invention and of the manner and process of making and using it, and is required to be in such full, clear, concise, and exact terms as to enable any person skilled in the technological area to which the invention pertains, or with which it is most nearly connected, to make and use the same.

The specification must set forth the precise invention for which a patent is solicited, in such manner as to distinguish it from other inventions and from what is old. It must describe completely a specific embodiment of the process, machine, manufacture, composition of matter, or improvement invented, and must explain the mode of operation or principle whenever applicable. The best mode contemplated by the inventor for carrying out the invention must be set forth.

In the case of an improvement, the specification must particularly point out the part or parts of the process, machine, manufacture, or composition of matter to which the improvement relates, and the description should be confined to the specific improvement and to such parts as necessarily cooperate with it or as may be necessary to a complete understanding or description of it.

The title of the invention, which should be as short and specific as possible (no more than 500 characters), should appear as a heading on the first page of the specification if it does not otherwise appear at the beginning of the application. A brief abstract of the technical disclosure in the specification, including that which is new in the art to which the invention pertains, must be set forth on a separate page preferably following the claims. The abstract should be in the form of a single paragraph of 150 words or less, under the heading "Abstract of the Disclosure."

A brief summary of the invention indicating its nature and substance, which may include a statement of the object of the invention, should precede the detailed description. The summary should be commensurate with the invention as claimed, and any object recited should be that of the invention as claimed.

When there are drawings, there shall be a brief description of the several views of the drawings, and the detailed description of the invention shall refer to the different views by specifying the numbers of the figures, and to the different parts by use of reference numerals.

The specification must conclude with a claim or claims particularly pointing out and distinctly claiming the subject matter that the applicant regards as the invention. The portion of the application in which the applicant sets forth the claim or claims is an important part of the application, as it is the claims that define the scope of the protection afforded by the patent and which questions of infringement are judged by the courts.

More than one claim may be presented, provided they differ substantially from each other and are not unduly multiplied. One or more claims may be presented in dependent form, referring back to and further limiting another claim or claims in the same application. Any dependent claim that refers back to more than one other claim is considered a "multiple dependent claim."

Multiple dependent claims shall refer to such other claims in the alternative only. A multiple dependent claim shall not serve as a basis for any other multiple dependent claims. Claims in dependent form shall be construed to include all of the limitations of the claim incorporated by reference into the dependent claim. A multiple dependent claim shall be construed to incorporate all the limitations of each of the particular claims in relation to which it is being considered.

The claim or claims must conform to the invention as set forth in the remainder of the specification and the terms and phrases used in the claims must find clear support or antecedent basis in the description so that the meaning of the terms in the claims may be ascertainable by reference to the description.

# Drawing

## Drawing

The applicant for a patent will be required by law to furnish a drawing of the invention whenever the nature of the case requires a drawing to understand the invention. However, the Director may require a drawing where the nature of the subject matter admits of it; this drawing must be filed with the application. This includes practically all inventions except compositions of matter or processes, but a drawing may also be useful in the case of many processes.

http://www.uspto.gov/patents-getting-started/general-information-con

The drawing must show every feature of the invention specified in the claims, and is required by the Office rules to be in a particular form. The Office specifies the size of the sheet on which the drawing is made, the type of paper, the margins, and other details relating to the making of the drawing. The reason for specifying the standards in detail is that the drawings are printed and published in a uniform style when the patent issues and the drawings must also be such that they can be readily understood by persons using the patent descriptions.

The sheets of drawings should be numbered in consecutive Arabic numerals, starting with 1, within the sight (the usable surface). For regular nonprovisional utility applications, these "sheets" should be contained in an electronic document in PDF format filed with the other application documents via EFS-Web. These numbers, if present, must be placed in the middle of the top of the sheet, but not in the margin. The numbers can be placed on the right-hand side if the drawing extends too close to the middle of the top edge of the usable surface. The drawing sheet numbering must be clear and larger than the numbers used as reference characters to avoid confusion. The number of each sheet should be shown by two Arabic numerals placed on either side of an oblique line, with the first being the sheet number and the second being the total number of sheets of drawings, with no other marking.

Identifying indicia, if provided, should include the title of the invention, the inventor's name, the application number (if known), and docket number (if any). This information should be placed on the top margin of each sheet of drawings. No names or other identification will be permitted within the "sight" of the drawing. The name and telephone number of a person to call if the USPTO is unable to match the drawings to the proper application may also be provided.

## Standards for Drawings

(1) Drawings. There are two acceptable categories for presenting drawings in utility and design patent applications:

(a) Black ink. Black and white drawings are normally required. India ink, or its equivalent that secures solid black lines, must be used for drawings, or

(b) Color. On rare occasions, color drawings may be necessary as the only practical medium by which to disclose the subject matter sought to be patented in a utility or design patent application or the subject matter of a statutory invention registration. The color drawings must be of sufficient quality such that all details in the drawings are reproducible in black and white in the printed patent. Color drawings are not permitted in international applications (see PCT Rule 11.13), or in an application, or copy thereof, submitted under the Office electronic filing system.

The Office will accept color drawings in utility or design patent applications and statutory invention registrations only after granting a petition filed under this paragraph explaining why the color drawings are necessary. Any such petition must include the following:

(i) The fee set forth in § 1.17(h);

(ii) Three sets of color drawings; and

(iii) An amendment to the specification to insert (unless the specification contains or has been previously amended to contain) the following language as the first paragraph of the brief description of the drawings:

The patent or application file contains at least one drawing executed in color. Copies of this patent or patent application publication with color drawing(s) will be provided by the Office upon request and payment of the necessary fee.

(2) Photographs

(a) Black and white. Photographs, including photocopies of photographs, are not ordinarily permitted in utility and design patent applications. The Office will accept photographs in utility and design patent applications, however, if photographs are the only practicable medium for illustrating the claimed invention. For example, photographs or photomicrographs of electrophoresis gels, blots (e.g., immuno-logical, western, southern, and northern), autoradiographs, cell cultures (stained and unstained), histological tissue cross sections (stained and unstained), animals, plants, in vivo imaging, thin-layer chromatography plates, crystalline structures, and, in a design patent application, ornamental effects, are acceptable. If the subject matter of the application admits of illustration by a drawing, the examiner may require a drawing in place of the photograph. The photographs must be of sufficient quality so that all details in the photographs are reproducible in the printed patent.

(b)Color photographs. Color photographs will be accepted in utility and design patent applications if the conditions for accepting color drawings and black and white photographs have been satisfied. See paragraphs (a)(2) and (b)(1) of this section.

(3) Identification of drawings - Identifying indicia should be provided, and if provided, should include the title of the invention, inventor's name, and application number, or docket number (if any) if an application number has not been assigned to the application. If this information is provided, it must be placed on the front of each sheet within the top margin. Each drawing sheet submitted after the filing date of an application must be identified as either "Replacement Sheet" or "New Sheet" pursuant to § 1.121(d). If a marked-up copy of any amended drawing figure including annotations indicating the changes made is filed, such marked-up copy must be clearly labeled as "Annotated Sheet" pursuant to § 1.121(d)(1).

(4) Graphic forms in drawings - Chemical or mathematical formulae, tables, and waveforms may be submitted as drawings and are subject to the same requirements as drawings. Each chemical or mathematical formula must be labeled as a separate figure, using brackets when necessary, to show that information is properly integrated. Each group of waveforms must be presented as a single figure, using a common vertical axis with time extending along the horizontal axis. Each individual waveform discussed in the specification must be identified with a separate letter designation adjacent to the vertical axis.

(5) Margins - The sheets must not contain frames around the sight (i.e., the usable surface), but should have scan target points (i.e., cross-hairs) printed on two cattycorner margin corners. Each sheet must include a top margin of at least 2.5 cm (1 inch), a left side margin of at least 2.5 cm (1 inch), a right side margin of at least 1.5 cm (5/8 inch), and a bottom margin of at least 1 cm (3/8 inch), thereby leaving a sight no greater than 17 cm by 26.2 cm on 21 cm by 29.7 cm (DIN size A4) drawing sheets, and a sight no greater than 17.6 cm by 24.4 cm (6 15/16 by 9 5/8 inches) on 21.6 cm by 27.9 cm (8 1/2 by 11 inch) drawing sheets.

(6) Views - The drawing must contain as many views as necessary to show the invention. The views may be plan, elevation, section, or perspective views. Detail views of portions of

elements, on a larger scale if necessary, may also be used. All views of the drawing must be grouped together and arranged on the sheet(s) without wasting space, preferably in an upright position, clearly separated from one another, and must not be included in the sheets containing the specifications, claims, or abstract. Views must not be connected by projection lines and must not contain center lines. Waveforms of electrical signals may be connected by dashed lines to show the relative timing of the waveforms.

(a) Exploded views – Exploded views with the separated parts embraced by a bracket, to show the relationship or order of assembly of various parts are permissible. When an exploded view is shown in a figure that is on the same sheet as another figure, the exploded view should be placed in brackets.

(b) Partial views - When necessary, a view of a large machine or device in its entirety may be broken into partial views on a single sheet, or extended over several sheets if there is no loss in facility of understanding the view. Partial views drawn on separate sheets must always be capable of being linked edge to edge so that no partial view contains parts of another partial view. A smaller scale view should be included showing the whole formed by the partial views and indicating the positions of the parts shown. When a portion of a view is enlarged for magnification purposes, the view and the enlarged view must each be labeled as separate views.

(i) Where views on two or more sheets form, in effect, a single complete view, the views on the several sheets must be so arranged that the complete figure can be assembled without concealing any part of any of the views appearing on the various sheets.

(ii) A very long view may be divided into several parts placed one above the other on a single sheet. However, the relationship between the different parts must be clear and unambiguous.

(c) Sectional views. The plane upon which a sectional view is taken should be indicated on the view from which the section is cut by a broken line. The ends of the broken line should be designated by Arabic or Roman numerals corresponding to the view number of the sectional view, and should have arrows to indicate the direction of sight. Hatching must be used to indicate section portions of an object, and must be made by regularly spaced oblique parallel lines spaced sufficiently apart to enable the lines to be distinguished without difficulty. Hatching should not impede the clear reading of the reference characters and lead lines. If it is not possible to place reference characters outside the hatched area, the hatching may be broken off wherever reference characters are inserted. Hatching must be at a substantial angle to the surrounding axes or principal lines, preferably 45 degrees. A cross section must be set out and drawn to show all of the materials as they are shown in the view from which the cross section was taken. The parts in cross section must show proper material(s) by hatching with regularly spaced parallel oblique strokes, the space between strokes being chosen on the basis of the total area to be hatched. The various parts of a cross section of the same item should be hatched in the same manner and should accurately and graphically indicate the nature of the material(s) that is illustrated in cross section. The hatching of juxtaposed different elements must be angled in a different way. In the case of large areas, hatching may be confined to an edging drawn around the entire inside of the outline of the area to be hatched. Different types of hatching should have different conventional meanings as regards the nature of a material seen in cross section.

(d) Alternate position. A moved position may be shown by a broken line superimposed upon a suitable view if this can be done without crowding; otherwise, a separate view must be used for this purpose.

(e) Modified forms. Modified forms of construction must be shown in separate views.

(7) Arrangement of views - One view must not be placed upon another or within the outline of another. All views on the same sheet should stand in the same direction and, if possible, stand so that they can be read with the sheet held in an upright position. If views wider than the width of the sheet are necessary for the clearest illustration of the invention, the sheet may be turned on its side so that the top of the sheet, with the appropriate top margin to be used as the heading space, is on the right-hand side. Words must appear in a horizontal, left-to-right fashion when the page is either upright or turned so that the top becomes the right side, except for graphs utilizing standard scientific convention to denote the axis of abscissas (of X) and the axis of ordinates (of Y).

(8) Front page view - The drawing must contain as many views as necessary to show the invention. One of the views should be suitable for inclusion on the front page of the patent application publication and patent as the illustration of the invention. Views must not be connected by projection lines and must not contain center lines. Applicant may suggest a single view (by figure number) for inclusion on the front page of the patent application publication and patent.

(9) Scale - The scale to which a drawing is made must be large enough to show the mechanism without crowding when the drawing is reduced in size to two-thirds in reproduction. Indications such as "actual size" or "scale 1/2" on the drawings are not permitted since these lose their meaning with reproduction in a different format.

(10) Character of lines, numbers, and letters - All drawings must be made by a process that will give them satisfactory reproduction characteristics. Every line, number, and letter must be durable, clean, black (except for color drawings), sufficiently dense and dark, and uniformly thick and well defined. The weight of all lines and letters must be heavy enough to permit adequate reproduction. This requirement applies to all lines however fine, to shading, and to lines representing cut surfaces in sectional views. Lines and strokes of different thicknesses may be used in the same drawing where different thicknesses have a different meaning.

(11) Shading - The use of shading in views is encouraged if it aids in understanding the invention and if it does not reduce legibility. Shading is used to indicate the surface or shape of spherical, cylindrical, and conical elements of an object. Flat parts may also be lightly shaded. Such shading is preferred in the case of parts shown in perspective, but not for cross sections. See paragraph (h)(3) of this section. Spaced lines for shading are preferred. These lines must be thin, as few in number as practicable, and they must contrast with the rest of the drawings. As a substitute for shading, heavy lines on the shade side of objects can be used except where they superimpose on each other or obscure reference characters. Light should come from the upper left corner at an angle of 45 degrees. Surface delineations should preferably be shown by proper shading. Solid black shading areas are not permitted, except when used to represent bar graphs or color.

(12) Symbols - Graphical drawing symbols may be used for conventional elements when appropriate. The elements for which such symbols and labeled representations are used must be adequately identified in the specification. Known devices should be illustrated by symbols that have a universally recognized conventional meaning and are generally accepted in the art. Other symbols which are not universally recognized may be used, subject to approval by the Office, if they are not likely to be confused with existing conventional symbols, and if they are readily identifiable.

(13) Legends - Suitable descriptive legends may be used subject to approval by the Office, or may be required by the examiner where necessary for understanding of the drawing. They should contain as few words as possible.

(14)Numbers, letters, and reference characters

(a) Reference characters (numerals are preferred), sheet numbers, and view numbers must be plain and legible, and must not be used in association with brackets or inverted commas, or enclosed within outlines, e.g., encircled. They must be oriented in the same direction as the view so as to avoid having to rotate the sheet. Reference characters should be arranged to follow the profile of the object depicted.

(b) The English alphabet must be used for letters, except where another alphabet is customarily used, such as the Greek alphabet to indicate angles, wavelengths, and mathematical formulas.

(c) Numbers, letters, and reference characters must measure at least .32 cm (1/8 inch) in height. They should not be placed in the drawing so as to interfere with its comprehension. Therefore, they should not cross or mingle with the lines. They should not be placed upon hatched or shaded surfaces. When necessary, such as indicating a surface or cross section, a reference character may be underlined and a blank space may be left in the hatching or shading where the character occurs so that it appears distinct.

(d) The same part of an invention appearing in more than one view of the drawing must always be designated by the same reference character, and the same reference character must never be used to designate different parts.

(e) Reference characters not mentioned in the description shall not appear in the drawings. Reference characters mentioned in the description must appear in the drawings.

(15) Lead lines - Lead lines are those lines between the reference characters and the details referred to. Such lines may be straight or curved and should be as short as possible. They must originate in the immediate proximity of the reference character and extend to the feature indicated. Lead lines must not cross each other. Lead lines are required for each reference character except for those which indicate the surface or cross section on which they are placed. Such a reference character must be underlined to make it clear that a lead line has not been left out by mistake. Lead lines must be executed in the same way as lines in the drawing. See paragraph (1) of this section.

(16) Arrows - Arrows may be used at the ends of lines, provided that their meaning is clear, as follows:

(a) On a lead line, a freestanding arrow to indicate the entire section towards which it points;

(b) On a lead line, an arrow touching a line to indicate the surface shown by the line looking along the direction of the arrow

(c) To show the direction of movement.

(17) Copyright or Mask Work Notice - A copyright or mask work notice may appear in the drawing, but must be placed within the sight of the drawing immediately below the figure representing the copyright or mask work material and be limited to letters having a print size of .32 cm. to .64 cm. (1/8 to 1/4 inches) high. The content of the notice must be limited to only those elements provided for by law. For example, "©1983 John Doe" (17 U.S.C. 401) and "*M* John Doe" (17 U.S.C. 909) would be properly limited and, under current statutes, legally sufficient notices of copyright and mask work, respectively. Inclusion of a copyright or mask work notice will be permitted only if the authorization language set forth in 1.71(e) is included at the beginning (preferably as the first paragraph) of the specification.

(18)Numbering of sheets of drawings - The sheets of drawings should be numbered in consecutive Arabic numerals, starting with 1, within the sight as defined in paragraph (5) of this section. These numbers, if present, must be placed in the middle of the top of the sheet, but not in the margin. The numbers can be placed on the right-hand side if the drawing extends too close to the middle of the top edge of the usable surface. The drawing sheet numbering must be clear and larger than the numbers used as reference characters to avoid confusion. The number of each sheet should be shown by two Arabic numerals placed on either side of an oblique line, with the first being the sheet number and the second being the total number of sheets of drawings, with no other marking.

(19) Numbering of views

(a) The different views must be numbered in consecutive Arabic numerals, starting with 1, independent of the numbering of the sheets and, if possible, in the order in which they appear on the drawing sheet(s). Partial views intended to form one complete view, on one or several sheets must be identified by the same number followed by a capital letter. View numbers must be preceded by the abbreviation "FIG." Where only a single view is used in an application to illustrate the claimed invention, it must not be numbered and the abbreviation "FIG." must not appear.

(b) Numbers and letters identifying the views must be simple and clear and must not be used in association with brackets, circles, or inverted commas. The view numbers must be larger than the numbers used for reference characters.

(20) Security markings - Authorized security markings may be placed on the drawings provided they are outside the sight, preferably centered in the top margin.

(21) Corrections - Any corrections on drawings submitted to the Office must be durable and permanent.

(22) Holes - No holes should be made by applicant in the drawing sheets.

(23) Types of drawings - See § 1.152 for design drawings, § 1.165 for plant drawings, and § 1.173(a)(2) for reissue drawings

## Models, Exhibits, and Specimens

Models or exhibits are not required in most patent applications since the description of the invention in the specification and the drawings must be sufficiently full, clear, and complete and capable of being understood to disclose the invention without the aid of a model.

A working model, or other physical exhibit, may be required by the Office if deemed necessary. This is not done very often. A working model may be requested in the case of applications for patent for alleged perpetual motion devices.

When the invention relates to a composition of matter, the applicant may be required to furnish specimens of the composition, or of its ingredients or intermediates, for inspection or experiment. If the invention is a microbiological invention, a deposit of the microorganism involved is required.

# Examination of Applications and Proceedings in the USPTO

Applications, other than provisional applications, filed in the United States Patent and Trademark Office and accepted as complete applications are assigned for examination to the respective examining technology centers having charge of the areas of technology related to the invention. In the examining TC, applications are taken up for examination by the examiner to whom they have been assigned in the order in which they have been filed or in accordance with examining procedures established by the Director.

Applications will not be advanced out of turn for examination or for further action except as provided by the rules, or upon order of the Director to expedite the business of the Office, or upon a showing that, in the opinion of the Director, will justify advancing them.

The examination of the application consists of a study of the application for compliance with the legal requirements and a search through U.S. patents, publications of patent applications, foreign patent documents, and available literature, to see if the claimed invention is new, useful and non-obvious and if the application meets the requirements of the patent statute and rules of practice. If the examiner's decision on patentability is favorable, a patent is granted.

## Restrictions

If two or more inventions are claimed in a single application, and are regarded by the Office to be of such a nature (e.g. independent and distinct) that a single patent should not be issued for both of them, the applicant will be required to limit the application to one of the inventions. The other invention may be made the subject of a separate application which, if filed while the first application is still pending, will be entitled to the benefit of the filing date of the first application. A requirement to restrict the application to one invention may be made before further action by the examiner.

## Office Action

The applicant is notified in writing of the examiner's decision by an Office "action" which is normally mailed to the attorney or agent of record. The reasons for any adverse action or any objection or requirement are stated in the Office action and such information or references are given as may be useful in aiding the applicant to judge the propriety of continuing the prosecution of his or her application.

If the claimed invention is not directed to patentable subject matter, the claims will be rejected. If the examiner finds that the claimed invention lacks novelty or differs only in an obvious manner from what is found in the prior art, the claims may also be rejected. It is not uncommon for some or all of the claims to be rejected on the first Office action by the examiner; relatively few applications are allowed as filed.

## Applicant's Reply

The applicant must request reconsideration in writing, and must distinctly and specifically point out the supposed errors in the examiner's Office action. The applicant must reply to every ground of objection and rejection in the prior Office action. The applicant's reply must appear throughout to be a bona fide attempt to advance the case to final action or allowance. The mere allegation that the examiner has erred will not be received as a proper reason for such reconsideration.

In amending an application in reply to a rejection, the applicant must clearly point out why he or she thinks the amended claims are patentable in view of the state of the art disclosed by the prior references cited or the objections made. He or she must also show how the claims as amended avoid such references or objections. After reply by the applicant, the application will be reconsidered, and the applicant will be notified as to the status of the claims—that is, whether the claims are rejected, or objected to, or whether the claims are allowed, in the same manner as after the first examination. The second Office action usually will be made final.

Interviews with examiners may be arranged, but an interview does not remove the necessity of replying to Office actions within the required time.

## Final Rejection

On the second or later consideration, the rejection or other action may be made final. The applicant's reply is then limited to appeal in the case of rejection of any claim and further amendment is restricted. Petition may be taken to the Director in the case of objections or requirements not involved in the rejection of any claim. Reply to a final rejection or action must include cancellation of, or appeal from the rejection of, each claim so rejected and, if any claim stands allowed, compliance with any requirement or objection as to form. In making such final rejection, the examiner repeats or states all grounds of rejection then considered applicable to the claims in the application.

## Amendments to Application

The applicant may amend the application as specified in the rules, or when and as specifically required by the examiner.

Amendments received in the Office on or before the mail date of the first Office action are called "preliminary amendments," and their entry is governed by 37 CFR 1.115. Amendments in reply to a non-final Office action are governed by CFR 1.111. Amendments filed after final action are governed by 37CFR 1.116 and 37CFR 41.33.

The specification, claims, and drawing must be amended and revised when required, to correct inaccuracies of description and definition or unnecessary words, and to provide substantial correspondence between the claims, the description, and the drawing. All amendments of the drawings or specification, and all additions thereto must not include new matter beyond the original disclosure. Matter not found in either, involving a departure from or an addition to the original disclosure cannot be added to the application even if supported by a supplemental oath or declaration, and can be shown or claimed only in a separate application.

The manner of making amendments to an application is provided in 37 CFR 1.121. Amendments to the specification (but not including the claims) must be made by adding, deleting or replacing a paragraph, by replacing a section, or by a substitute specification, as provided in the rules. Replacement paragraphs are to include markings (e.g., underlining and strikethrough) to show all changes relative to the previous version of the paragraph. New paragraphs are to be provided without any underlining. If a substitute specification is filed, it must be submitted with markings (e.g., underlining and strikethrough) showing all the changes relative to the immediate prior version of the specification of record, it must be accompanied by a statement that the substitute specification includes no new matter, and it must be accompanied by a clean version without markings.

No change in the drawing may be made except by permission of the Office. Changes in the construction shown in any drawing may be made only by submitting replacement drawing sheets, each of which must be labeled "Replacement Sheet" in its top margin if it replaces an existing drawing sheet. Any replacement sheet of drawings must include all of the figures appearing on the immediate prior version of the sheet, even if only one figure is amended. Any new sheet of drawings containing an additional figure must be labeled in the top margin as "New Sheet." All changes to the drawings must be explained, in detail, in either the drawing amendment or remarks section of the amendment paper.

Amendments to the claims are to be made by presenting all of the claims in a claim listing that replaces all prior versions of the claims in the application. In the claim listing, the status of every claim must be indicated after its claim number after using one of the seven parenthetical expressions set forth in 37 CFR 1.121(c). "Currently amended" claims must be submitted with markings (e.g., underlining and strikethrough). All pending claims not being currently amended must be presented in the claim listing in clean version without any markings (e.g., underlining and strikethrough).

The original numbering of the claims must be preserved throughout the prosecution. When claims are canceled, the remaining claims must not be renumbered. When claims are added by amendment or substituted for canceled claims, they must be numbered by the applicant consecutively beginning with the number next following the highest numbered claim previously presented. When the application is ready for allowance, the examiner, if necessary, will renumber the claims consecutively in the order in which they appear or in such order as may have been requested by applicant.

## Time for Reply and Abandonment

The reply of an applicant to an action by the Office must be made within a prescribed time limit. The maximum period for reply is set at six months by the statute (35 U.S.C. 133), which also provides that the Director may shorten the time for reply to not less than 30 days. The usual period for reply to an Office action is three months. A shortened time for reply may be extended up to the maximum six-month period. An extension of time fee is normally required to be paid if the reply period is extended. The amount of the fee is dependent upon the length of the extension. Extensions of time are generally not available after an application has been allowed. If no reply is received within the time period, the application is considered as abandoned and no longer pending. However, if it can be shown that the failure to prosecute was unavoidable or unintentional, the application may be revived upon request to and approval by the Director. The revival requires a petition to the Director, and a fee for the petition, which must be filed without delay. The proper reply must also accompany the petition if it has not yet been filed.

http://www.uspto.gov/patents-getting-started/general-information-con

## Allowance and Issue of Patent

If, on examination of the application, or at a later stage during the reconsideration of the application, the patent application is found to be allowable, a Notice of Allowance and Fee(s) Due will be sent to the applicant, or to applicant's attorney or agent of record, if any, and a fee for issuing the patent and if applicable, for publishing the patent application publication (see 37 CFR 1.211-1.221), is due within three months from the date of the notice. If timely payment of the fee(s) is not made, the application will be regarded as abandoned. See the current fee schedule at www.uspto.gov. The Director may accept the fee(s) late, if the delay is shown to be unavoidable (35 U.S.C. 41, 37 CFR 1.137(a)) or unintentional (35 U.S.C. 151, 37 CFR 1.137(b)). When the required fees are paid, the patent issues as soon as possible after the date of payment, dependent upon the volume of printing on hand. The patent grant then is delivered or mailed on the day of its grant, or as soon thereafter as possible, to the inventor's attorney or agent if there is one of record, otherwise directly to the inventor. On the date of the grant, the patent file becomes open to the public for applications not opened earlier by publication of the application.

In cases where the publication of an application or the granting of a patent would be detrimental to the national security, the Commissioner for Patents will order that the invention be kept secret and shall withhold the publication of the application or the grant of the patent for such period as the national interest requires. The owner of an application that has been placed under a secrecy order has a right to appeal the order to the Secretary of Commerce. 35 U.S.C. 181.

## Patent Term Extension and Adjustment

The terms of certain patents may be subject to extension or adjustment under 35 U.S.C. 154(b). Such extension or adjustment results from certain specified types of delays which may occur while an application is pending before the Office.

Utility and plant patents which issue from original applications filed between June 8, 1995 and May 28, 2000 may be eligible for patent term extension (PTE) as set forth in 37 CFR 1.701. Such PTE may result from delays due to interference proceedings under 35 U.S.C. 135(a), secrecy orders under 35 U.S.C. 181, or successful appellate review.

Utility and plant patents which issue from original applications filed on or after May 29, 2000 may be eligible for patent term adjustment (PTA) as set forth in 37 CFR 1.702 - 1.705. There are three main bases for PTA under 35 U.S.C. 154(b). The first basis for PTA is the failure of the Office to take certain actions within specific time frames set forth in 35 U.S.C. 154(b)(1)(A) (See 37 CFR 1.702(a) and 1.703(a)). The second basis for PTA is the failure of the Office to issue a patent within three years of the actual filing date of the application as set forth in 35 U.S.C. 154(b)(1)(B) (See 37 CFR 1.702(b) and 1.703(b)). The third basis for PTA is set forth in 35 U.S.C. 154(b)(1)(C), and includes delays due to interference proceedings under 35 U.S.C. 135(a), secrecy orders under 35 U.S.C. 181, or successful appellate review (See 37 CFR 1.702(c)-(e) and 1.703(c)-(e)).

Any PTA which has accrued in an application will be reduced by the time period during which an applicant failed to engage in reasonable efforts to conclude prosecution of the application pursuant to 35 U.S.C. 154(b)(2)(C). A non-exclusive list of activities which constitute failure to engage in reasonable efforts to conclude prosecution is set forth in 37 CFR 1.704.24

An initial PTA value is printed on the notice of allowance and fee(s) due, and a final PTA value is printed on the front of the patent. Any request for reconsideration of the PTA value printed on the notice of allowance and fee(s) due should be made in the form of an application for patent term adjustment, which must be filed prior to or at the same time as the payment of the issue fee. (See 37 CFR 1.705.)

## Nature of Patent and Patent Rights

The patent is issued in the name of the United States under the seal of the United States Patent and Trademark Office, and is either signed by the Director of the USPTO or is electronically written thereon and attested by an Office official. The patent contains a grant to the patentee, and a printed copy of the specification and drawing is annexed to the patent and forms a part of it. The grant confers "the right to exclude others from making, using, offering for sale, or selling the invention throughout the United States or importing the invention into the United States" and its territories and possessions for which the term of the patent shall be generally 20 years from the date on which the application for the patent was filed in the United States or, if the application contains a specific reference to an earlier filed application under 35 U.S.C. 120, 121 or 365(c), from the date of the earliest such application was filed, and subject to the payment of maintenance fees as provided by law.

The exact nature of the right conferred must be carefully distinguished, and the key is in the words "right to exclude" in the phrase just quoted. The patent does not grant the right to make, use, offer for sale or sell or import the invention but only grants the exclusive nature of the right. Any person is ordinarily free to make, use, offer for sale or sell or import anything he or she pleases, and a grant from the government is not necessary. The patent only grants the right to exclude others from making, using, offering for sale or selling or importing the invention. Since the patent does not grant the right to make, use, offer for sale, or sell, or import the invention, the patentee's own right to do so is dependent upon the rights of others and whatever general laws might be applicable. A patentee, merely because he or she has received a patent for an invention, is not thereby authorized to make, use, offer for sale, or sell, or import the invention if doing so would violate any law.

An inventor of a new automobile who has obtained a patent thereon would not be entitled to use the patented automobile in violation of the laws of a state requiring a license, nor may a patentee sell an article, the sale of which may be forbidden by a law, merely because a patent has been obtained.

Neither may a patentee make, use, offer for sale, or sell, or import his or her own invention if doing so would infringe the prior rights of others. A patentee may not violate the federal antitrust laws, such as by resale price agreements or entering into combination in restraints of trade, or the pure food and drug laws, by virtue of having a patent. Ordinarily there is nothing that prohibits a patentee from making, using, offering for sale, or selling, or importing his or her own invention, unless he or she thereby infringes another's patent that is still in force. For example, a patent for an improvement of an original device already patented would be subject to the patent on the device.

The term of the patent shall be generally 20 years from the date on which the application for the patent was filed in the United States or, if the application contains a specific reference to an earlier filed application under 35 U.S.C. 120, 121 or 365(c), from the date of the earliest such application was filed, and subject to the payment of maintenance fees as provided by law. A maintenance fee is due 3.5, 7.5 and 11.5 years after the original grant for all patents issuing from the applications filed on and after December 12, 1980. The maintenance fee must be paid at the stipulated times to maintain the patent in force. After the patent has expired anyone may make, use, offer for sale, or sell or import the invention without permission of the patentee, provided that matter covered by other unexpired patents is not used. The terms may be extended for certain pharmaceuticals and for certain circumstances as provided by law.

## Maintenance Fees

All utility patents that issue from applications filed on or after December 12, 1980 are subject to the payment of maintenance fees which must be paid to maintain the patent in force. These fees are due at 3.5, 7.5 and 11.5 years from the date the patent is granted and can be paid without a surcharge during the "window period," which is the six-month period preceding each due date, e.g., three years to three years and six months. (See fee schedule for a list of maintenance fees.) In submitting maintenance fees and any necessary surcharges, identification of the patents for which maintenance fees are being paid must include the patent number, and the application number of the U.S. application for the patent on which the maintenance fee is being paid. If the payment includes identification of only the patent number, the Office may apply payment to the patent identified by patent number in the payment or the Office may return the payment. (See 37, Code of Federal Regulations, section 1.366(c).)

Failure to pay the current maintenance fee on time may result in expiration of the patent. A six-month grace period is provided when the maintenance fee may be paid with a surcharge. The grace period is the six-month period immediately following the due date. The USPTO does not mail notices to patent owners that maintenance fees are due. If, however, the maintenance fee is not paid on time, efforts are made to remind the responsible party that the maintenance fee may be paid during the grace period with a surcharge. If the maintenance fee is not paid on time and the maintenance fee and surcharge are not paid during the grace period, the patent expires on the date the grace period ends.

## Correction of Patents

Once the patent is granted, it is outside the jurisdiction of the USPTO except in a few respects. The Office may issue without charge a certificate correcting a clerical error it has made in the patent when the printed patent does not correspond to the record in the Office. These are mostly corrections of typographical errors made in printing. Some minor errors of a typographical nature made by the applicant may be corrected by a certificate of correction for which a fee is required. The patentee may disclaim one or more claims of his or her patent by filing in the Office a disclaimer as provided by the statute (35 U.S.C. 253).

When the patent is defective in certain respects, the law provides that the patentee may apply for a reissue patent. Following an examination in which the proposed changes correcting any defects in the original patent are evaluated, a reissue patent would be granted to replace the original and is granted only for the balance of the unexpired term. However, the nature of the changes that can be made by means of the reissue are rather limited; new matter cannot be added. In a different type of proceeding, any person may file a request for reexamination of a patent, along with the required fee, on the basis of prior art consisting of patents or printed publications. At the conclusion of the reexamination proceedings, a certificate setting forth the results of the reexamination proceeding is issued.

## Assignments and Licenses

A patent is personal property and may be sold to others or mortgaged; it may be bequeathed by a will; and it may pass to the heirs of a deceased patentee. The patent law provides for the transfer or sale of a patent, or of an application for patent, by an instrument in writing. Such an instrument is referred to as an assignment and may transfer the entire interest in the patent. The assignee, when the patent is assigned to him or her, becomes the owner of the patent and has the same rights that the original patentee had.

The statute also provides for the assignment of a part interest, that is, a half interest, a fourth interest, etc., in a patent. There may also be a grant that conveys the same character of interest as an assignment but only for a particularly specified part of the United States. A mortgage of patent property passes ownership thereof to the mortgagee or lender until the mortgage has been satisfied and a retransfer from the mortgagee back to the mortgagor, the borrower, is made. A conditional assignment also passes ownership of the patent and is regarded as absolute until canceled by the parties or by the decree of a competent court.

An assignment, grant, or conveyance of any patent or application for patent should be acknowledged before a notary public or officer authorized to administer oaths or perform notarial acts. The certificate of such acknowledgment constitutes prima facie evidence of the execution of the assignment, grant, or conveyance.

### Recording of Assignments

The Office records assignments, grants, and similar instruments sent to it for recording, and the recording serves as notice. If an assignment, grant, or conveyance of a patent or an interest in a patent (or an application for patent) is not recorded in the Office within three months from its date, it is void against a subsequent purchaser for a valuable consideration without notice, unless it is recorded prior to the subsequent purchase.

An instrument relating to a patent should identify the patent by number and date (the name of the inventor and title of the invention as stated in the patent should also be given). An instrument relating to an application should identify the application by its application number and date of filing, the name of the inventor, and title of the invention as stated in the application should also be given. Sometimes an assignment of an application is executed at the same time that the application is prepared and before it has been filed in the Office. Such assignment should adequately identify the application, as by its date of execution and name of the inventor and title of the invention, so that there can be no mistake as to the application intended. If an application has been assigned and the assignment has been recorded or filed for recordation, the patent will be issued to the assignee as owner, if the name of the assignee is provided when the issue fee is paid and the patent is requested to be issued to the assignee. If the assignment is of a part interest only, the patent will be issued to the inventor and assignee as joint owners.

**Joint Ownership**

Patents may be owned jointly by two or more persons as in the case of a patent granted to joint inventors, or in the case of the assignment of a part interest in a patent. Any joint owner of a patent, no matter how small the part interest, may make, use, offer for sale and sell and import the invention for his or her own profit provided they do not infringe another's patent rights, without regard to the other owners, and may sell the interest or any part of it, or grant licenses to others, without regard to the other joint owner, unless the joint owners have made a contract governing their relation to each other. It is accordingly dangerous to assign a part interest without a definite agreement between the parties as to the extent of their respective rights and their obligations to each other if the above result is to be avoided.

The owner of a patent may grant licenses to others. Since the patentee has the right to exclude others from making, using, offering for sale, or selling or importing the invention, no one else may do any of these things without his or her permission.

A patent license agreement is in essence nothing more than a promise by the licensor not to sue the licensee. No particular form of license is required; a license is a contract and may include whatever provisions the parties agree upon, including the payment of royalties, etc.

The drawing up of a license agreement (as well as assignments) is within the field of an attorney at law. Such an attorney should be familiar with patent matters as well. A few states have prescribed certain formalities to be observed in connection with the sale of patent rights.

## Infringement of Patents

Infringement of a patent consists of the unauthorized making, using, offering for sale, or selling any patented invention within the United States or U.S. Territories, or importing into the United States of any patented invention during the term of the patent. If a patent is infringed, the patentee may sue for relief in the appropriate federal court. The patentee may ask the court for an injunction to prevent the continuation of the infringement and may also ask the court for an award of damages because of the infringement. In such an infringement suit, the defendant may raise the question of the validity of the patent, which is then decided by the court. The defendant may also aver that what is being done does not constitute infringement. Infringement is determined primarily by the language of the claims of the patent and, if what the defendant is making does not fall within the language of any of the claims of the patent, there is no literal infringement.

Suits for infringement of patents follow the rules of procedure of the federal courts. From the decision of the district court, there is an appeal to the Court of Appeals for the Federal Circuit. The Supreme Court may thereafter take a case by writ of certiorari. If the United States Government infringes a patent, the patentee has a remedy for damages in the United States Court of Federal Claims. The government may use any patented invention without permission of the patentee, but the patentee is entitled to obtain compensation for the use by or for the government. The Office has no jurisdiction over questions relating to infringement of patents. In examining applications for patent, no determination is made as to whether the invention sought to be patented infringes any prior patent. An improvement invention may be patentable, but it might infringe a prior unexpired patent for the invention improved upon, if there is one.

## Patent Marking and Patent Pending

A patentee who makes or sells patented articles, or a person who does so for or under the patentee is required to mark the articles with the word "patent" and the number of the patent. The penalty for failure to mark is that the patentee may not recover damages from an infringer unless the infringer was duly notified of the infringement and continued to infringe after the notice.

The marking of an article as patented when it is not in fact patented is against the law and subjects the offender to a penalty. Some persons mark articles sold with the terms "Patent Applied For" or "Patent Pending." These phrases have no legal effect, but only give information that an application for patent has been filed in the USPTO. The protection afforded by a patent does not start until the actual grant of the patent. False use of these phrases or their equivalent is prohibited.

# Part 3 – Filing patent application - Provisional

## Provisional Application

Since June 8, 1995, the USPTO has offered inventors the option of filing a provisional application for patent, which was designed to provide a lower-cost first patent filing in the United States and to give U.S. applicants parity with foreign applicants. Claims and oath or declaration are NOT required for a provisional application. A provisional application provides the means to establish an early effective filing date in a patent application and permits the term "Patent Pending" to be applied in connection with the invention. Provisional applications may not be filed for design inventions.

The filing date of a provisional application is the date on which a written description of the invention, and drawings if necessary, are received in the USPTO. To be complete, a provisional application must also include the filing fee, and a cover sheet specifying that the application is a provisional application for patent. The applicant would then have up to 12 months to file a nonprovisional application for patent as described above. The claimed subject matter in the later filed nonprovisional application is entitled to the benefit of the filing date of the provisional application if it has support in the provisional application.

If a provisional application is not filed in English, and a nonprovisional application is filed claiming benefit to the provisional application, a translation of the provisional application will be required. See title 37, Code of Federal Regulations, Section 1.78(a)(5).

Provisional applications are NOT examined on their merits. A provisional application will become abandoned by the operation of law 12 months from its filing date. The 12-month pendency for a provisional application is not counted toward the 20-year term of a patent granted on a subsequently filed nonprovisional application that claims benefit of the filing date of the provisional application.

A surcharge is required for filing the basic filing fee or the cover sheet on a date later than the filing of the provisional application. Unlike nonprovisional utility applications, design, plant, and provisional applications can still be filed by mail or hand-delivery without having to pay the additional $400 non-electronic filing fee. Design and provisional applications can also be filed via EFS-Web. Plant applications, however, are not permitted to be filed via EFS-Web.

# Cover Page

## Cover Sheet Page 1

Degree of difficulty of filling highlighted items: EASY

Doc Code: TR.PROV
Document Description: Provisional Cover Sheet (SB16)

PTO/SB/16 (11-08)
Approved for use through 05/31/2015. OMB 0651-0032
U.S. Patent and Trademark Office; U.S. DEPARTMENT OF COMMERCE
Under the Paperwork Reduction Act of 1995, no persons are required to respond to a collection of information unless it displays a valid OMB control number

### Provisional Application for Patent Cover Sheet

This is a request for filing a PROVISIONAL APPLICATION FOR PATENT under 37 CFR 1.53(c)

**Inventor(s)**

Inventor 1                                                                    [ Remove ]

| Given Name | Middle Name | Family Name | City | State | Country |
|---|---|---|---|---|---|
| Roy | | Lique | Walnut | Ca | US |

All Inventors Must Be Listed – Additional Inventor Information blocks may be generated within this form by selecting the **Add** button.   [ Add ]

| Title of Invention | Digital Camera Lens Guard and Use Extender |
|---|---|
| Attorney Docket Number (if applicable) | |

**Correspondence Address**

Direct all correspondence to (select one):

○ The address corresponding to Customer Number     ◉ Firm or Individual Name

| Firm or Individual Name 1 | Roy  Lique |
|---|---|
| Firm or Individual Name 2 | |

**Mailing Address of Applicant:**

| Address 1 | 1431 Tierra Cima | | |
|---|---|---|---|
| Address 2 | | | |
| City | Walnut | State/Province | CA |
| Postal Code | 91789 | Country | US |
| Phone | 9095941470 | | |

The invention was made by an agency of the United States Government or under a contract with an agency of the United States Government.

◉ No.
○ Yes, the name of the U.S. Government agency and the Government contract number are:

EFS - Web 1.0.1

## Cover Sheet Page 2

Degree of
difficulty of
filling
highlighted
items: EASY

Duc Code: **TR.PROV**
Document Description: Provisional Cover Sheet (SB16)

PT0/SB/16 (11-08)
Approved for use through 05/31/2015. OMB 0651-0032
U.S. Patent and Trademark Office: U.S. DEPARTMENT OF COMMERCE
Under the Paperwork Reduction Act of 1995, no persons are required to respond to a collection of information unless it displays a valid OMB control number

**Entity Status**
Applicant claims small entity status under 37 CFR 1.27

◉ Yes, applicant qualifies for small entity status under 37 CFR 1.27
○ No

**Warning**

Petitioner/applicant is cautioned to avoid submitting personal information in documents filed in a patent application that may contribute to identity theft. Personal information such as social security numbers, bank account numbers, or credit card numbers (other than a check or credit card authorization form PTO-2038 submitted for payment purposes) is never required by the USPTO to support a petition or an application. If this type of personal information is included in documents submitted to the USPTO, petitioners/applicants should consider redacting such personal information from the documents before submitting them to USPTO. Petitioner/applicant is advised that the record of a patent application is available to the public after publication of the application (unless a non-publication request in compliance with 37 CFR 1.213(a) is made in the application) or issuance of a patent. Furthermore, the record from an abandoned application may also be available to the public if the application is referenced in a published application or an issued patent (see 37 CFR1.14). Checks and credit card authorization forms PTO-2038 submitted for payment purposes are not retained in the application file and therefore are not publicly available.

**Signature**

Please see 37 CFR 1.4(d) for the form of the signature.

| Signature | | | Date (YYYY-MM-DD) | 2013-01-21 |
|---|---|---|---|---|
| First Name | Roy | Last Name | Lique | Registration Number (if appropriate) | |

This collection of information is required by 37 CFR 1.51. The information is required to obtain or retain a benefit by the public which is to file (and by the USPTO to process) an application. Confidentiality is governed by 35 U.S.C. 122 and 37 CFR 1.11 and 1.14. This collection is estimated to take 8 hours to complete, including gathering, preparing, and submitting the completed application form to the USPTO. Time will vary depending upon the individual case. Any comments on the amount of time you require to complete this form and/or suggestions for reducing this burden, should be sent to the Chief Information Officer, U.S. Patent and Trademark Office, U.S. Department of Commerce, P.O. Box 1450, Alexandria, VA 22313-1450. DO NOT SEND FEES OR COMPLETED FORMS TO THIS ADDRESS. This form can only be used when in conjunction with EFS-Web. If this form is mailed to the USPTO, it may cause delays in handling the provisional application.

EFS - Web 1.0.1

# Acknowledgement

## Acknowledgement – Page 1

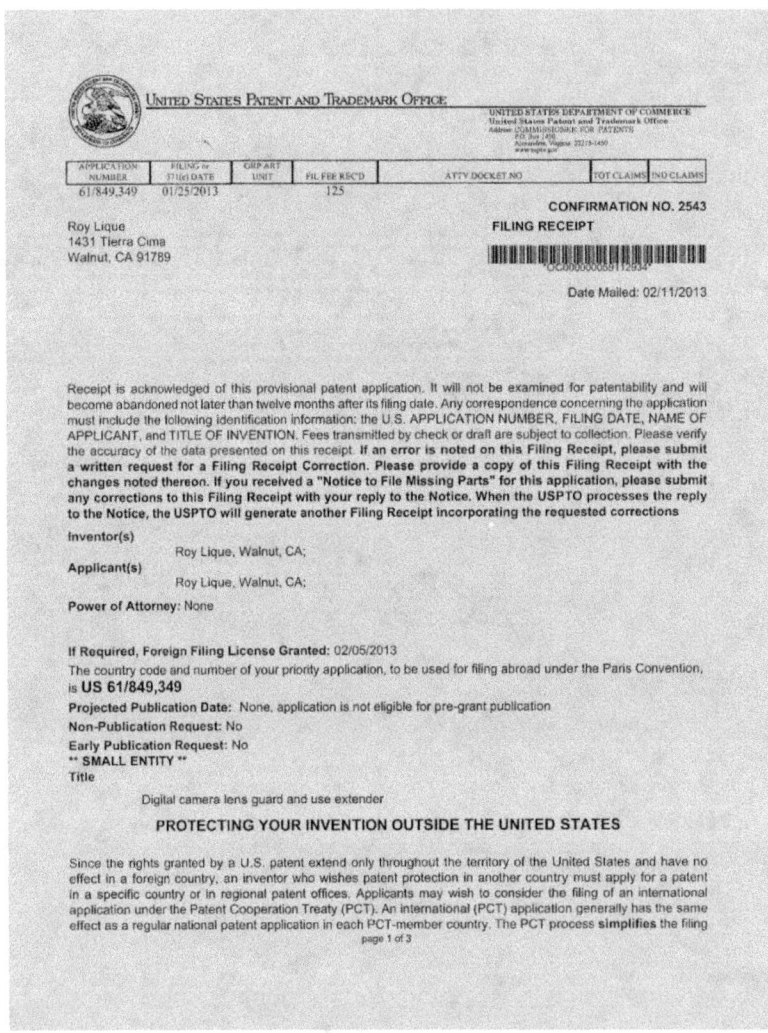

## Acknowledgement – Page 2

of patent applications on the same invention in member countries, but **does not result** in a grant of "an international patent" and does not eliminate the need of applicants to file additional documents and fees in countries where patent protection is desired.

Almost every country has its own patent law, and a person desiring a patent in a particular country must make an application for patent in that country in accordance with its particular laws. Since the laws of many countries differ in various respects from the patent law of the United States, applicants are advised to seek guidance from specific foreign countries to ensure that patent rights are not lost prematurely.

Applicants also are advised that in the case of inventions made in the United States, the Director of the USPTO must issue a license before applicants can apply for a patent in a foreign country. The filing of a U.S. patent application serves as a request for a foreign filing license. The application's filing receipt contains further information and guidance as to the status of applicant's license for foreign filing.

Applicants may wish to consult the USPTO booklet, "General Information Concerning Patents" (specifically, the section entitled "Treaties and Foreign Patents") for more information on timeframes and deadlines for filing foreign patent applications. The guide is available either by contacting the USPTO Contact Center at 800-786-9199, or it can be viewed on the USPTO website at http://www.uspto.gov/web/offices/pac/doc/general/index.html.

For information on preventing theft of your intellectual property (patents, trademarks and copyrights), you may wish to consult the U.S. Government website, http://www.stopfakes.gov. Part of a Department of Commerce initiative, this website includes self-help "toolkits" giving innovators guidance on how to protect intellectual property in specific countries such as China, Korea and Mexico. For questions regarding patent enforcement issues, applicants may call the U.S. Government hotline at 1-866-999-HALT (1-866-999-4158).

### LICENSE FOR FOREIGN FILING UNDER

#### Title 35, United States Code, Section 184

#### Title 37, Code of Federal Regulations, 5.11 & 5.15

**GRANTED**

The applicant has been granted a license under 35 U.S.C. 184, if the phrase "IF REQUIRED, FOREIGN FILING LICENSE GRANTED" followed by a date appears on this form. Such licenses are issued in all applications where the conditions for issuance of a license have been met, regardless of whether or not a license may be required as set forth in 37 CFR 5.15. The scope and limitations of this license are set forth in 37 CFR 5.15(a) unless an earlier license has been issued under 37 CFR 5.15(b). The license is subject to revocation upon written notification. The date indicated is the effective date of the license, unless an earlier license of similar scope has been granted under 37 CFR 5.13 or 5.14.

This license is to be retained by the licensee and may be used at any time on or after the effective date thereof unless it is revoked. This license is automatically transferred to any related applications(s) filed under 37 CFR 1.53(d). This license is not retroactive.

The grant of a license does not in any way lessen the responsibility of a licensee for the security of the subject matter as imposed by any Government contract or the provisions of existing laws relating to espionage and the national security or the export of technical data. Licensees should apprise themselves of current regulations especially with respect to certain countries, of other agencies, particularly the Office of Defense Trade Controls, Department of

page 2 of 3

# Acknowledgement – Page 3

State (with respect to Arms, Munitions and Implements of War (22 CFR 121-128)); the Bureau of Industry and Security, Department of Commerce (15 CFR parts 730-774); the Office of Foreign AssetsControl, Department of Treasury (31 CFR Parts 500+) and the Department of Energy.

**NOT GRANTED**

No license under 35 U.S.C. 184 has been granted at this time, if the phrase "IF REQUIRED, FOREIGN FILING LICENSE GRANTED" DOES NOT appear on this form. Applicant may still petition for a license under 37 CFR 5.12, if a license is desired before the expiration of 6 months from the filing date of the application. If 6 months has lapsed from the filing date of this application and the licensee has not received any indication of a secrecy order under 35 U.S.C. 181, the licensee may foreign file the application pursuant to 37 CFR 5.15(b).

---

## *SelectUSA*

The United States represents the largest, most dynamic marketplace in the world and is an unparalleled location for business investment, innovation, and commercialization of new technologies. The U.S. offers tremendous resources and advantages for those who invest and manufacture goods here. Through SelectUSA, our nation works to promote and facilitate business investment. SelectUSA provides information assistance to the international investor community; serves as an ombudsman for existing and potential investors; advocates on behalf of U.S. cities, states, and regions competing for global investment; and counsels U.S. economic development organizations on investment attraction best practices. To learn more about why the United States is the best country in the world to develop technology, manufacture products, deliver services, and grow your business, visit http://www.SelectUSA.gov or call +1-202-482-6800.

page 3 of 3

## Application Data Sheet

## Application Data Sheet - Page 1

Degree of difficulty of filling highlighted items: EASY

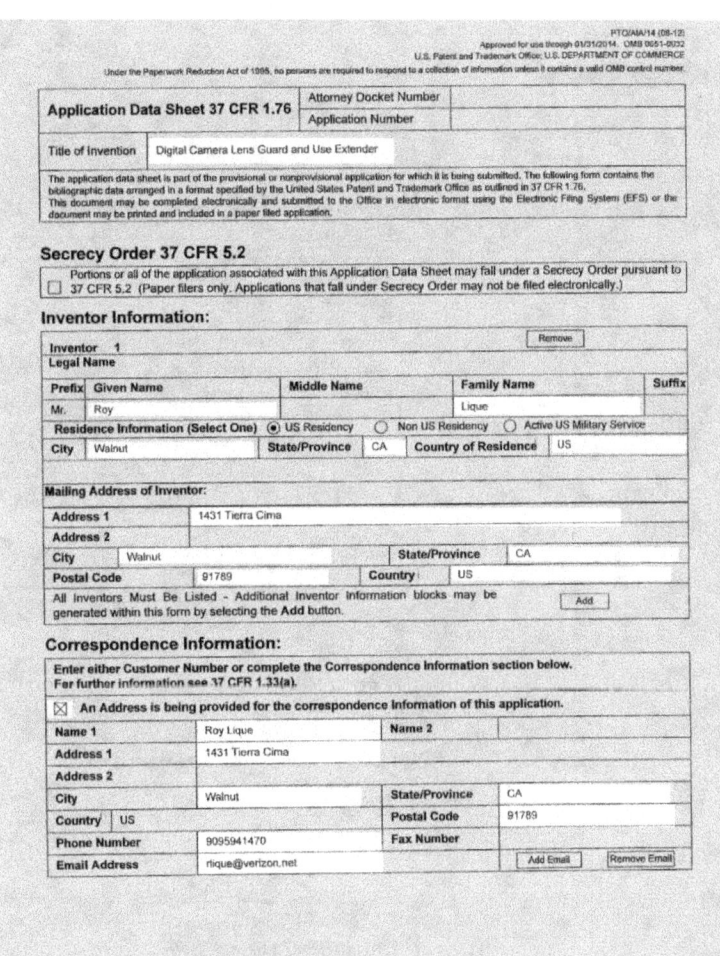

## Application Data Sheet - Page 2

**Degree of difficulty of filling highlighted items: EASY**

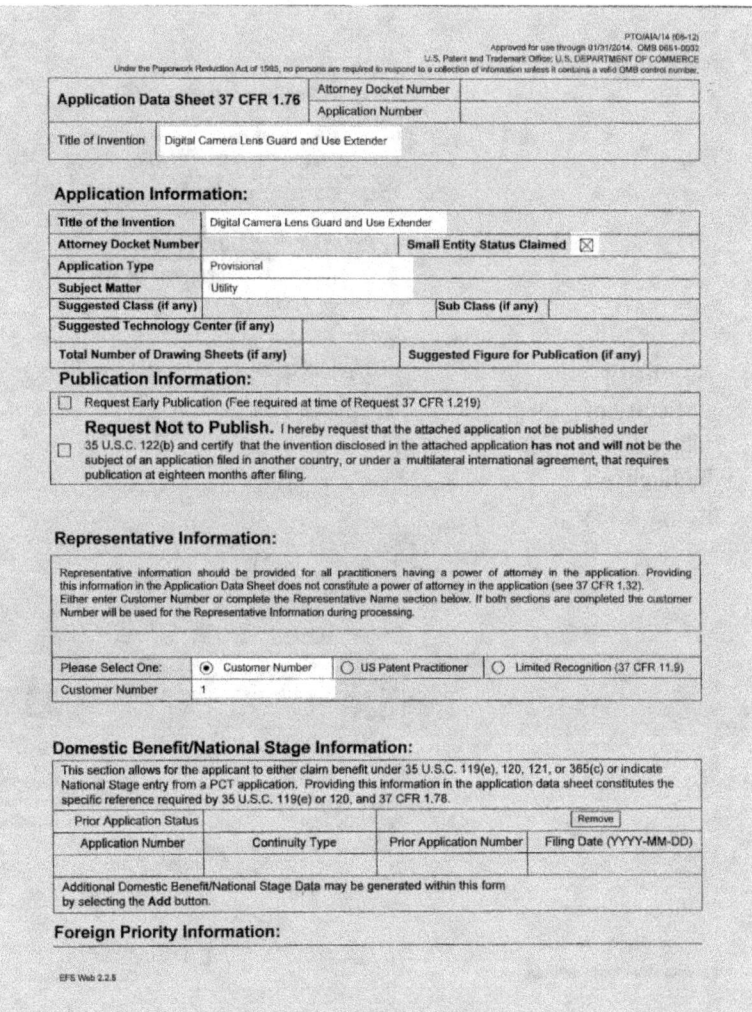

## Application Data Sheet - Page 3

**Degree of difficulty of filling highlighted items: EASY**

PTO/AIA/14 (08-12)
Approved for use through 01/31/2014. OMB 0651-0032
U.S. Patent and Trademark Office. U.S. DEPARTMENT OF COMMERCE
Under the Paperwork Reduction Act of 1995, no persons are required to respond to a collection of information unless it contains a valid OMB control number.

| Application Data Sheet 37 CFR 1.76 | Attorney Docket Number | |
| | Application Number | |

| Title of Invention | Digital Camera Lens Guard and Use Extender |

This section allows for the applicant to claim benefit of foreign priority and to identify any prior foreign application for which priority is not claimed. Providing this information in the application data sheet constitutes the claim for priority as required by 35 U.S.C. 119(b) and 37 CFR 1.55(a).

Remove

| Application Number | Country | Filing Date (YYYY-MM-DD) | Priority Claimed |
| | | | ○ Yes   ○ No |

Additional Foreign Priority Data may be generated within this form by selecting the **Add** button.

### Authorization to Permit Access:

☐   Authorization to Permit Access to the Instant Application by the Participating Offices

If checked, the undersigned hereby grants the USPTO authority to provide the European Patent Office (EPO), the Japan Patent Office (JPO), the Korean Intellectual Property Office (KIPO), the World Intellectual Property Office (WIPO), and any other intellectual property offices in which a foreign application claiming priority to the instant patent application is filed access to the instant patent application. See 37 CFR 1.14(c) and (h). This box should not be checked if the applicant does not wish the EPO, JPO, KIPO, WIPO, or other intellectual property office in which a foreign application claiming priority to the instant patent application is filed to have access to the instant patent application.

In accordance with 37 CFR 1.14(h)(3), access will be provided to a copy of the instant patent application with respect to: 1) the instant patent application-as-filed; 2) any foreign application to which the instant patent application claims priority under 35 U.S.C. 119(a)-(d) if a copy of the foreign application that satisfies the certified copy requirement of 37 CFR 1.55 has been filed in the instant patent application; and 3) any U.S. application-as-filed from which benefit is sought in the instant patent application.

In accordance with 37 CFR 1.14(c), access may be provided to information concerning the date of filing this Authorization.

### Applicant Information:

Providing assignment information in this section does not substitute for compliance with any requirement of part 3 of Title 37 of CFR to have an assignment recorded by the Office.

EFS Web 2.2.9

## Application Data Sheet - Page 4

**Degree of difficulty of filling highlighted items: EASY**

PTO/AIA/14 (09-12)
Approved for use through 01/31/2016. OMB 0651-0032
U.S. Patent and Trademark Office; U.S. DEPARTMENT OF COMMERCE
Under the Paperwork Reduction Act of 1995, no persons are required to respond to a collection of information unless it contains a valid OMB control number.

| Application Data Sheet 37 CFR 1.76 | Attorney Docket Number | |
|---|---|---|
| | Application Number | |

| Title of Invention | Digital Camera Lens Guard and Use Extender |
|---|---|

**Applicant 1**

If the applicant is the inventor (or the remaining joint inventor or inventors under 37 CFR 1.45), this section should not be completed. The information to be provided in this section is the name and address of the legal representative who is the applicant under 37 CFR 1.43; or the name and address of the assignee, person to whom the inventor is under an obligation to assign the invention, or person who otherwise shows sufficient proprietary interest in the matter who is the applicant under 37 CFR 1.46. If the applicant is an applicant under 37 CFR 1.46 (assignee, person to whom the inventor is obligated to assign, or person who otherwise shows sufficient proprietary interest) together with one or more joint inventors, then the joint inventor or inventors who are also the applicant should be identified in this section.

[ Clear ]

| ○ Assignee | ○ Legal Representative under 35 U.S.C. 117 | ○ Joint Inventor |
|---|---|---|
| ○ Person to whom the inventor is obligated to assign. | ○ Person who shows sufficient proprietary interest | |

If applicant is the legal representative, indicate the authority to file the patent application, the inventor is:

Name of the Deceased or Legally Incapacitated Inventor :

If the Applicant is an Organization check here. ☐

| Prefix | Given Name | Middle Name | Family Name | Suffix |
|---|---|---|---|---|
| | | | | |

**Mailing Address Information:**

| Address 1 | |
|---|---|
| Address 2 | |
| City | State/Province |
| Country | Postal Code |
| Phone Number | Fax Number |
| Email Address | |

Additional Applicant Data may be generated within this form by selecting the Add button.

**Non-Applicant Assignee Information:**

Providing assignment information in this section does not subsitute for compliance with any requirement of part 3 of Title 37 of CFR to have an assignment recorded by the Office.

**Assignee 1**

Complete this section only if non-applicant assignee information is desired to be included on the patent application publication in accordance with 37 CFR 1.215(b). Do not include in this section an applicant under 37 CFR 1.46 (assignee, person to whom the inventor is obligated to assign, or person who otherwise shows sufficient proprietary interest), as the patent application publication will include the name of the applicant(s).

EFS Web 2.2.5

## Application Data Sheet - Page 5

Degree of
difficulty of
filling
highlighted
items: EASY

| Application Data Sheet 37 CFR 1.76 | Attorney Docket Number | |
| | Application Number | |

PTO/AIA/14 (05-12)
Approved for use through 01/31/2014. OMB 0651-0032
U.S. Patent and Trademark Office: U.S. DEPARTMENT OF COMMERCE
Under the Paperwork Reduction Act of 1995, no persons are required to respond to a collection of information unless it contains a valid OMB control number.

| Title of Invention | Digital Camera Lens Guard and Use Extender |

If the Assignee is an Organization check here. ☐

| Prefix | Given Name | Middle Name | Family Name | Suffix |
|--------|-----------|-------------|-------------|--------|
| | | | | |

**Mailing Address Information:**

| Address 1 | |
| Address 2 | |

| City | | State/Province | |
| Country | | Postal Code | |
| Phone Number | | Fax Number | |
| Email Address | |

Additional Assignee Data may be generated within this form by selecting the Add button.

**Signature:**

NOTE: This form must be signed in accordance with 37 CFR 1.33. See 37 CFR 1.4 for signature requirements and certifications

| Signature | *Roy Lique* | | Date (YYYY-MM-DD) | 2013-01-21 |
| First Name | Roy | Last Name | Lique | Registration Number | |

Additional Signature may be generated within this form by selecting the Add button.

This collection of information is required by 37 CFR 1.76. The information is required to obtain or retain a benefit by the public which is to file (and by the USPTO to process) an application. Confidentiality is governed by 35 U.S.C. 122 and 37 CFR 1.14. This collection is estimated to take 23 minutes to complete, including gathering, preparing, and submitting the completed application data sheet form to the USPTO. Time will vary depending upon the individual case. Any comments on the amount of time you require to complete this form and/or suggestions for reducing this burden, should be sent to the Chief Information Officer, U.S. Patent and Trademark Office, U.S. Department of Commerce, P.O. Box 1450, Alexandria, VA 22313-1450. DO NOT SEND FEES OR COMPLETED FORMS TO THIS ADDRESS. **SEND TO: Commissioner for Patents, P.O. Box 1450, Alexandria, VA 22313-1450.**

EFS Web 2.2.5

## Specification

## Specification – Page 1

### Digital Camera Lens Guard and Use Extender

#### BACKGROUND OF THE INVENTION

The present invention is in the field of digital cameras. Specifically, the present invention deals with protecting the lens of certain models of cameras and extending their uses and functionalities.

Camera ownership is increasing significantly. With the growing free times, people are commemorating more events with cameras. Pictures are attached to emails. They are uploaded to the internet. Businesses use images for presentations and researches. Interviews and news casting require extensive use of cameras. These activities create the need for a camera to be ready in an instant. More importantly, the camera must be specially equipped to be able to shoot clear images of objects a few inches away and of objects that are hundreds to thousands of feet away.

#### SUMMARY OF THE INVENTION

This invention protects the lens of certain models of digital cameras at the same time that it enhances and extends the cameras' functions. This is made possible by coupling the cameras to sighting devices.

#### BRIEF DESCRIPTIONS OF THE PHOTOGRAPHS

Fig. 1 is a perspective view of the major parts of the invention.
Fig. 2 is a perspective view of the base of the invention attached to the camera.
Fig. 3 is a perspective view of the main housing ring of the invention with drilled holes.
Fig. 4 is a perspective inside view of the invention as viewed from the front.
Fig. 5 is a perspective inside view of the invention as viewed from the rear.
Fig. 6 is a perspective view of the assembled main housing of the invention.
Fig. 7 is a perspective view of the invention attached to the camera and sighting device.

#### DETAILED DESCRIPTION OF THE INVENTION

Referring now to the invention in more detail, in Fig. 1 there is shown the unassembled major parts of the invention: base anchor ring 10, base retaining nut 12, extension tube 14, coupler stop 16, main housing ring 18, coupler ring 20, coupler retaining screw 22, and base retaining screw 24.

Referring now to the invention in more detail, in Fig. 2 there is shown the base anchor ring 10, with base retaining nut 12, and base retaining screw 24, attached to a digital camera.

Referring now to the invention in more detail, in Fig. 3 there is shown the main housing ring 18, with optional holes for future expansion and enhancements.

1

## Specification – Page 2

Referring now to the invention in more detail, in Fig. 4 there is shown the partially assembled invention with coupler stop 16, main housing ring 18, coupler retaining screw 22, and extension tube 14.

Referring now to the invention in more detail, in Fig. 5 there is shown the rear view of the invention with extension tube 14, main housing ring 18, coupler stop 16, and coupler retaining screws 22.

Referring now to the invention in more detail, in Fig. 6 there is shown the assembled main housing of the invention with extension tube 14, main housing ring 18, and coupler retaining screws 22.

Referring now to the invention in more detail, in Fig. 7 there is shown the invention attached to a digital camera and optional sighting device with base anchor ring 10, base retaining nut 12, base retaining screw 24, main housing ring 18, coupler retaining screws 22, and coupler ring 20.

In more detail, referring to the invention in Fig. 2, the base anchor ring 10, is attached to the digital camera using commercially available adhesives. When inserted, the assembled main housing is held at the base anchor ring 10, by tightening the base retaining screw 24, inserted through the base retaining nut 12.

In more detail, referring to the invention in Fig. 3 the main housing ring 18, is predrilled for inserting coupler retaining screws 22, and optional future expansions and enhancements.

In more detail, referring to the invention in Fig. 4, coupler retaining screws 22, are diametrically screwed into the main housing ring 18. The coupler stop 16, is anchored where it meets the extension tube 14, deeper into the main housing ring 18.

In more detail, referring to the invention in Fig. 5, coupler retaining screws 22, are screwed diametrically into the main housing ring 18. Using commercially available adhesives or solder, the extension tube 14, is permanently and partly anchored into the main housing ring 18, leaving enough protrusion to match the height of the base anchor ring 10 (not shown). Using commercially available adhesives or solder, the coupler stop 16, is anchored where the extension tube 14, ends in the main housing ring 18.

In more detail, referring to the invention in Fig. 6, the coupler retaining screws 22, are screwed diametrically into the main housing ring 18. Using commercially available adhesives or solder the extension tube 14, is permanently and partly anchored into the main housing ring 18, leaving enough protrusion to match the height of the base anchor ring 10 (not shown).

In more detail, referring to the invention in Fig. 7, the fully assembled Digital Camera Lens Guard and Use Extender showing the base anchor ring 10, base retaining nut 12, base retaining screw 24, main housing ring 18, coupler retaining screws 22, and the coupler ring 20, is attached to a digital camera on one end and to a sighting device on the other end.

2

**Specification – Page 3**

Referring to the invention in Fig. 1, Fig. 2, Fig. 3, Fig. 4, Fig. 5, Fig. 6, and Fig.7, the construction details of the invention are that the parts may be made of wood or any other sufficiently strong materials such as high-strength plastic, metal and the like. Further, the various components of the invention may be made of different materials.

The benefits of the present invention include, but not limited to the ability to protect the lens of certain models of digital cameras without limiting their original functions. Digital camera's functionalities are enhanced by their ability to be coupled with sighting devices thereby adding some of the latter's features to the cameras.

In broad embodiment, the present invention is an enhancement and an addition to the functionalities of certain models of digital cameras.

While the foregoing written description of the invention enables one of ordinary skill to make and use what is considered presently to be the best mode thereof, those of ordinary skills will understand and appreciate the existence of variations, combinations, and equivalents of the specific embodiment, method, and examples herein. The invention should therefore not be limited by the above described embodiment, method, and examples, but by all embodiments and methods within the scope and spirit of the invention as claimed.

3

**Specification – Page 4**

CLAIM OR CLAIMS

4

**Specification – Page 5**

ABSTRACT OF THE DISCLOSURE

A Digital Camera Lens Guard and Use Extender.

## Drawings

## Drawing – Figure 1

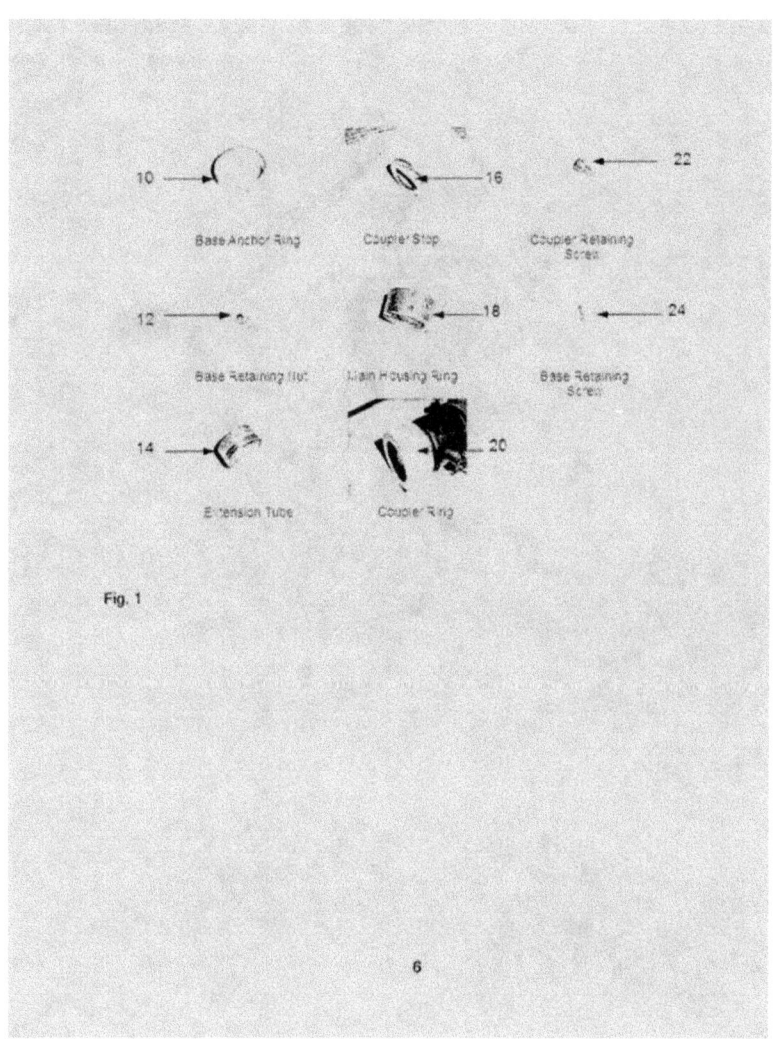

Fig. 1

## Drawing – Figure 2

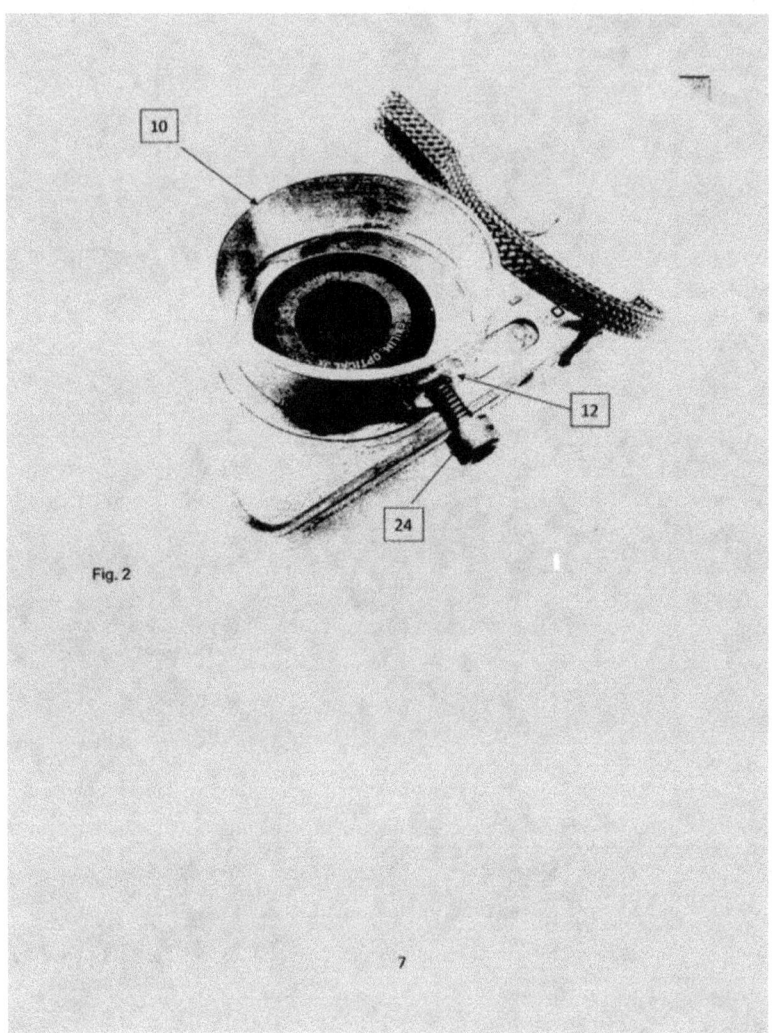

Fig. 2

7

**Drawing – Figure 3**

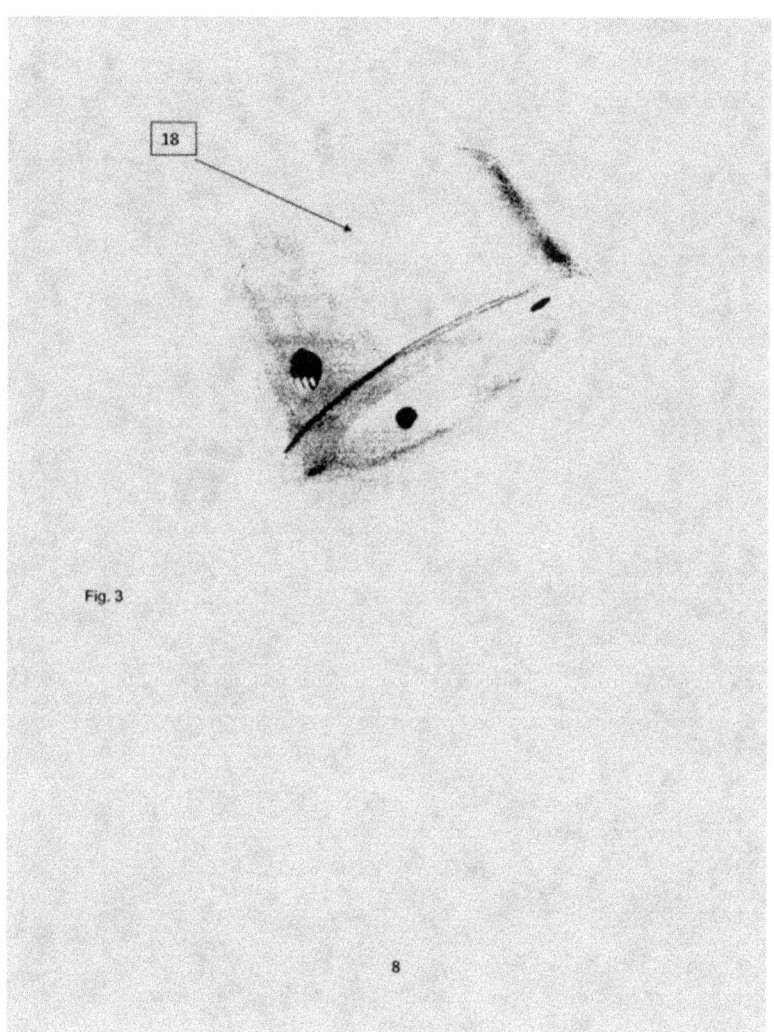

Fig. 3

8

**Drawing – Figure 4**

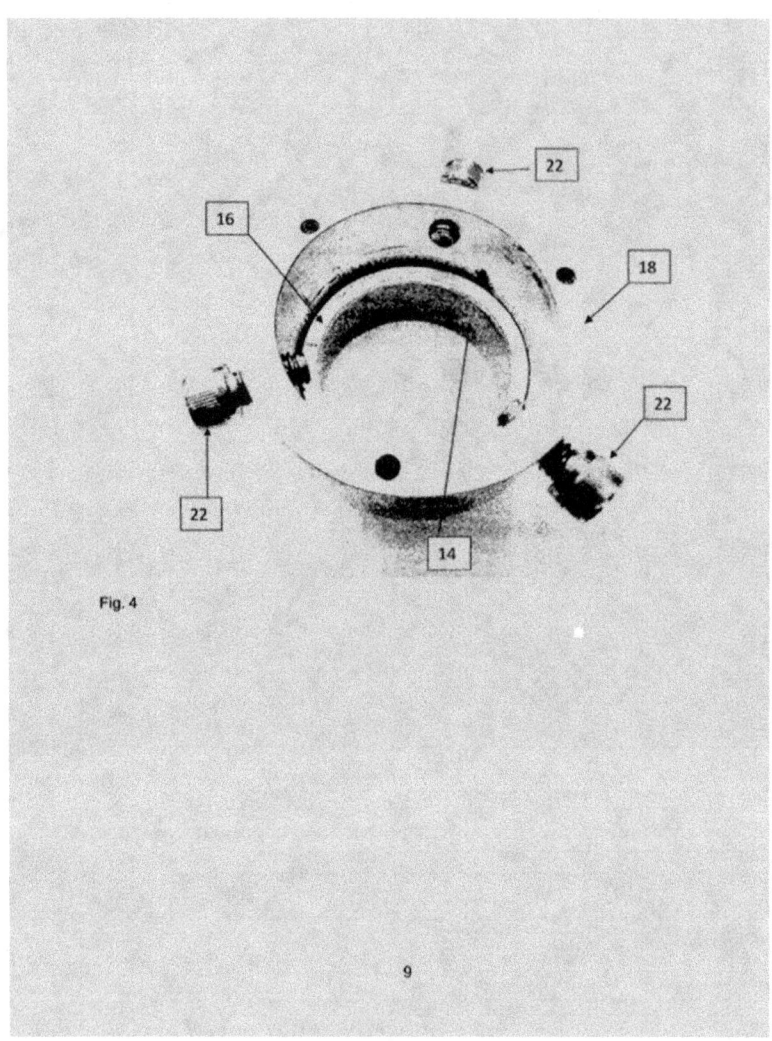

Fig. 4

9

## Drawing – Figure 5

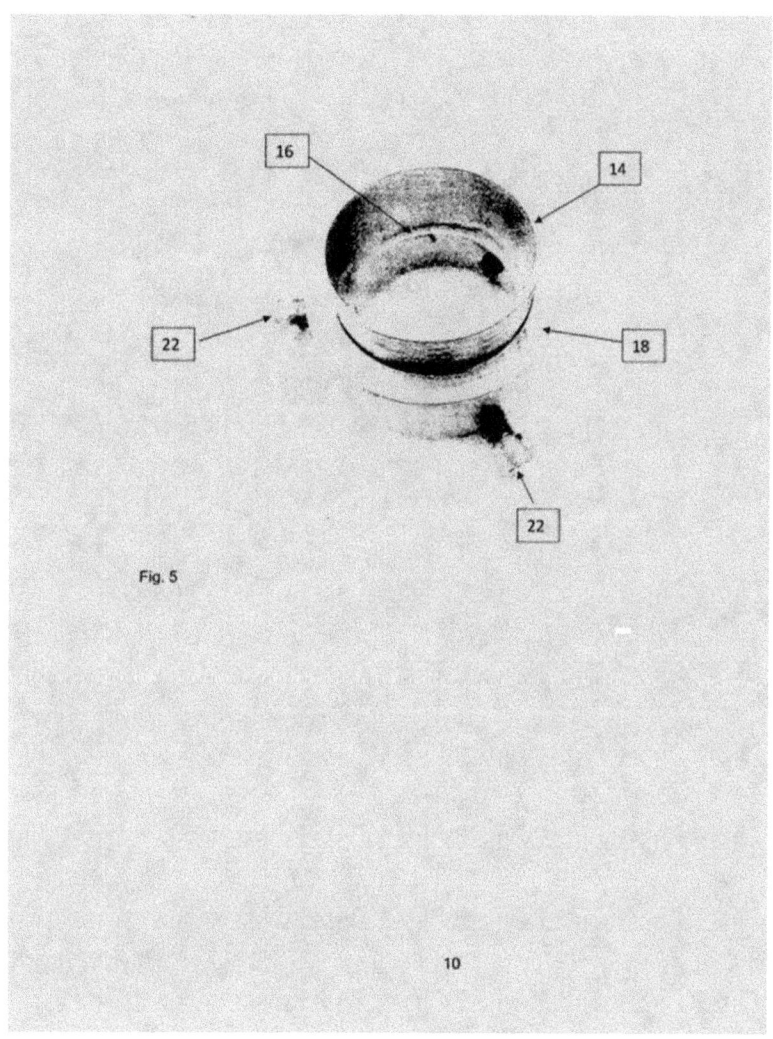

Fig. 5

10

**Drawing – Figure 6**

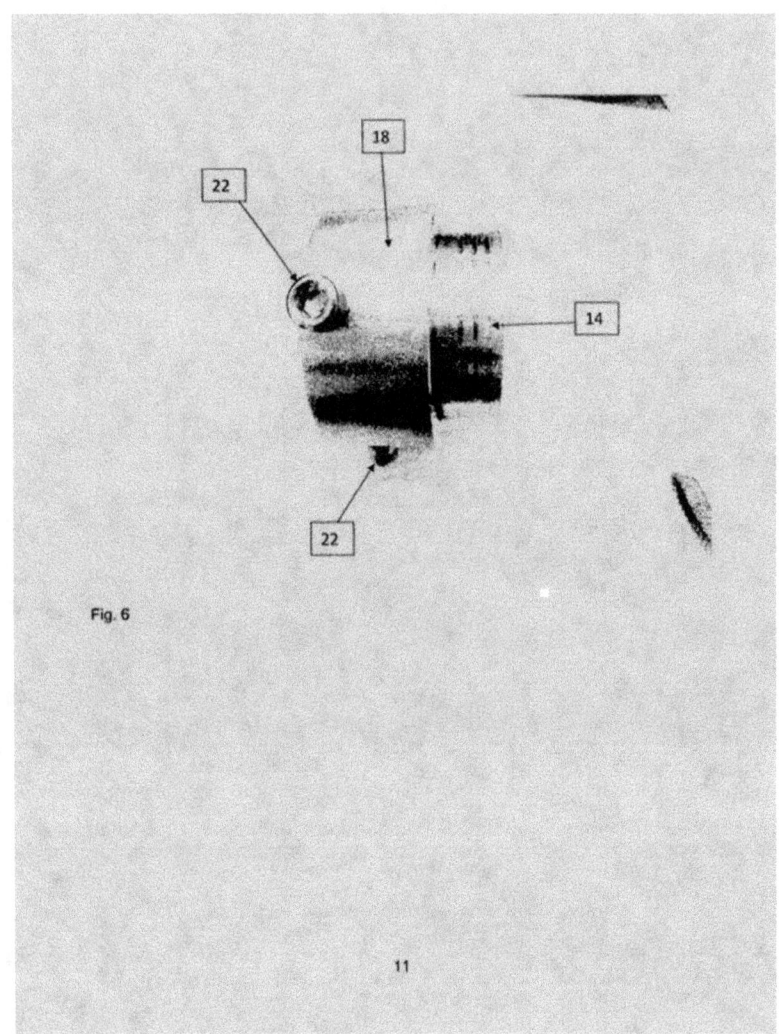

Fig. 6

**Drawing – Figure 7**

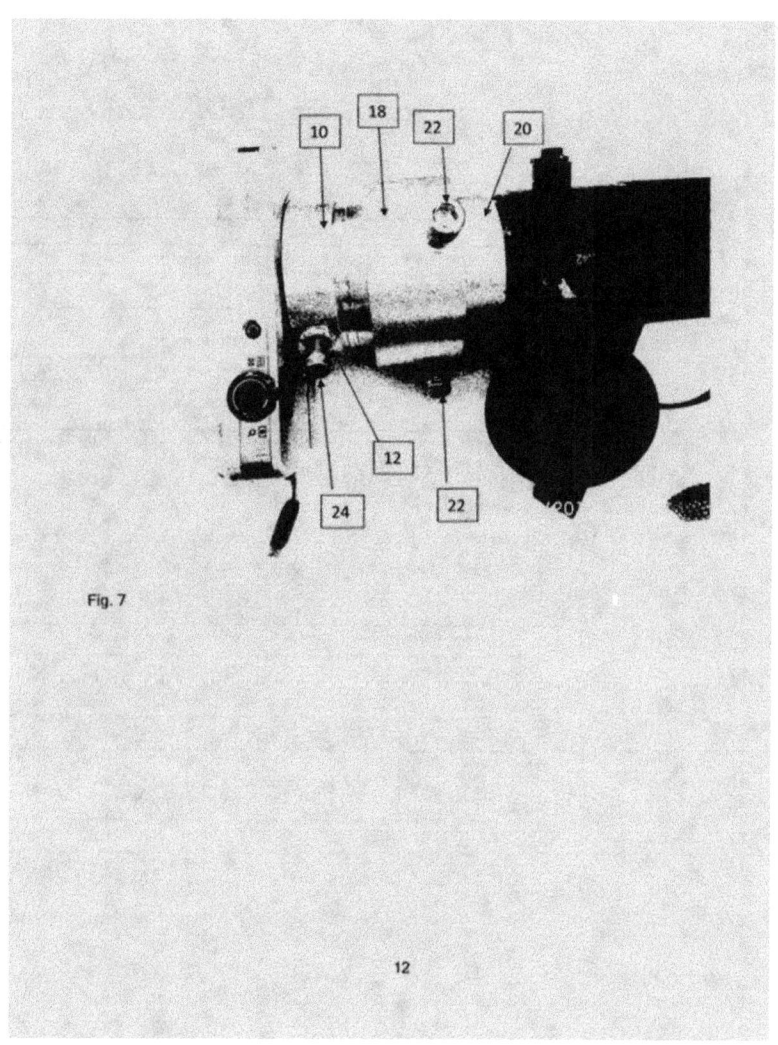

Fig. 7

12

# Part 4 – Patent Application Information Retrieval

## PAIR

**Request for Customer Number**

Degree of
difficulty of
filling
highlighted
items: EASY

PTO/SB/125A (11-08)
Approved for use through 11/30/2011. OMB 0651-0035
U.S. Patent and Trademark Office. U.S. DEPARTMENT OF COMMERCE
Under the Paperwork Reduction Act of 1995, no persons are required to respond to a collection of information unless it displays a valid OMB control number.

**Request for**
**Customer Number**

**Address to:**
Mail Stop CN
Commissioner for Patents
P.O. Box 1450
Alexandria, VA 22313-1450

To the Commissioner for Patents
Please assign a Customer Number to the Address indicated below.

| | |
|---|---|
| Firm or Individual Name | Roy LIQUE |
| Address | 1431 Tierra Cima |
| City | Walnut | State | CA | Zip | 91789 |
| Country | United States |
| Telephone | 909-594-1470 | Email | rlique@verizon.net |

Please associate the following practitioner registration number(s) with the Customer Number assigned to the Address cited above.

☐ Additional practitioner registration numbers are listed on supplemental sheet(s) attached hereto.

**Request Submitted by:**

| | |
|---|---|
| Firm Name (if applicable) | |
| Signature | Roy lique |
| Name of person submitting request | Roy LIQUE | Date 01-20-2013 |
| Registration Number, if applicable | | Telephone Number 909-594-1470 |

This collection of information is required by 37 CFR 1.33. The information is required to obtain or retain a benefit by the public which is to file (and by the USPTO to process) an application. Confidentiality is governed by 35 U.S.C. 122 and 37 CFR 1.11 and 1.14. This collection is estimated to take 12 minutes to complete, including gathering, preparing, and submitting the completed application form to the USPTO. Time will vary depending upon the individual case. Any comments on the amount of time you require to complete this form and/or suggestions for reducing this burden, should be sent to the Chief Information Officer, U.S. Patent and Trademark Office, U.S. Department of Commerce, P.O. Box 1450, Alexandria, VA 22313-1450. DO NOT SEND FEES OR COMPLETED FORMS TO THIS ADDRESS. SEND TO: Mail Stop CN, Commissioner for Patents, P.O. Box 1450, Alexandria, VA 22313-1450.

If you need assistance in completing the form, call 1-800-PTO-9199 (1-800-786-9199) and select option 2.

## Certificate Action Form – Page 1

Degree of
difficulty of
filling
highlighted
items: EASY

The form has to
be notarized

# Certificate Action Form – Page 2

Privacy
Statement

**Privacy Act Statement**

This information is collected under the authority of 35 USC 2 and 122. This information is also being collected in conjunction with the provisions of the Government Paperwork Elimination Act. This information will only be used by the U.S. Patent and Trademark Office (USPTO) staff to issue and revoke digital certificates and to recover keys. It is requested that you supply this information so that the USPTO can authorize the creation of a digital certificate. This digital certificate enables the USPTO to issue the cryptographic "keys" necessary to provide you with a digital identity and to support encrypted communication between you and the USPTO. This information will be used to construct a unique name (distinguished name) and to communicate with you about the certificate grant and software distribution process. Furnishing the information on this form is voluntary; but failure to do so may result in disapproval of this request.

The information provided by you in this form will be subject to the following routine uses:

1. This information may be disclosed to Federal, state, local, or foreign agencies responsible for investigating, prosecuting, enforcing, or implementing laws, contracts, rules, or regulations, if these records indicate a violation or a potential violation of a law or contract. These violations or potential violations can be civil, criminal, or regulatory in nature and can arise from general or particular program statutes or contracts, rules, regulations, or from the necessity of protecting an interest of the Department.

2. A record from this system of records may be disclosed to a Federal, state or local agency maintaining civil, criminal or other relevant enforcement information or other pertinent information, such as current licenses, if necessary to obtain information relevant to a Department decision concerning the assignment, hiring or retention of an individual, the issuance of a security clearance, the letting of a contract, or the issuance of a license, grant or other benefit.

3. A record from this system of records may be disclosed in the course of presenting evidence to a court, magistrate, or administrative tribunal, including disclosures to opposing counsel in the course of settlement negotiations.

4. A record in this system of records may be disclosed to a member of Congress or to a congressional staff member in response to an inquiry of the Congressional office made at the written request of the constituent about whom the record is maintained.

5. A record in this system of records may be disclosed to the Office of Management and Budget in connection with the review of private relief legislation (as set forth in OMB Circular No. A-19) at any stage of the legislative coordination and clearance process as described in the Circular.

6. A record in this system of records may be disclosed to the Department of Justice to determine whether disclosure is required by the Freedom of Information Act (FOIA).

7. The information may be disclosed to the agency contractors, grantees, experts, consultants, or volunteers who have been engaged by the agency to assist in the performance of a service related to this system of records and who have need to have access to the records in order to perform the activity. Recipients of information shall be required to comply with the requirements of the Privacy Act of 1974, as amended, pursuant to 5 USC 552a(m).

8. The information may be disclosed to the Office of Personnel Management (OPM) for personnel research purposes as a data source for management information, for the production of summary descriptive statistics and analytical studies in support of the function for which the records are collected and maintained, or for related man-power studies.

9. Records from this system of records may be disclosed to the National Archives and Records Administration or to the General Services Administration for records management inspections conducted under 44 USC §§ 2904 and 2906.

10. When a record on its face, or in conjunction with other records, indicates a violation or potential violation of law, whether it is civil, criminal, or regulatory in nature, and whether it arises from a general or particular program statute, a regulation, rule, or order, the record may be disclosed to the appropriate Federal, foreign, state, local or tribal agency, or to other public authorities responsible for enforcing, investigating, or prosecuting violations, or to those agencies charged with enforcing or implementing statutes, rules, regulations, or orders, if it is determined that the information is relevant to any enforcement, regulatory, investigative, or prosecutive responsibility of the receiving entity.

## PKI Request and Reference Number

When Request for Customer Number is in order and approved, a reference number is issued.

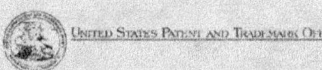

UNITED STATES PATENT AND TRADEMARK OFFICE

Roy Lique
1431 Tierra Cima
Walnut CA 91789

Your PKI request has been processed and your Reference Number issued.

**Reference # 30818193**

The PKI Certificate Reference Number must be used in conjunction with the Authorization Code that was emailed to you previously from our Security Program Office.

Now that you have both your Authorization Code and the Reference Number, you are able to create your digital certificate for access to PAIR and take advantage of electronic filing of patent applications or patent application status checks via the Internet. To create your digital certificate, go to the Digital Certificate Management page at the USPTO website: https://dcm.uspto.gov/UserRegistration/do/Home. As part of the process you will be required to input the Reference Number above and the Authorization Code provided to you previously by e-mail. You will also be asked to create a password for future access to your account.

Please note the Authorization Code lifespan is 100 days, starting on the date the code was e-mailed to you. At the end of the 100 days, it is set to expire if it has not been used to download software as described in the previous paragraph. If your code expires, please contact the Patent Electronic Business Center (EBC) to request reactivation of your Authorization Code.

Please review introductory and reference materials on the complete line of Patent electronic tools at http://www.uspto.gov/patentcenter/ebc/about.jsp under the Electronic Business Center heading. Our electronic tools allow on-line access for customers to submit patent applications using the Electronic Filing System (EFS), to view patent documents via Image File Wrapper (IFW), check the status of pending applications, and access full e-filing support.

If you have any questions or need additional information, the Patent EBC can be reached toll free at 866-217-9197 between the hours of 6 am and midnight Monday through Friday EST, by email at: ebc@uspto.gov or at our website at: http://www.uspto.gov/patents/ebc/about.jsp.

Sincerely,
Patent Electronic Business Center

## Customer Number Assignment

A unique customer number is assigned which only owner can use.

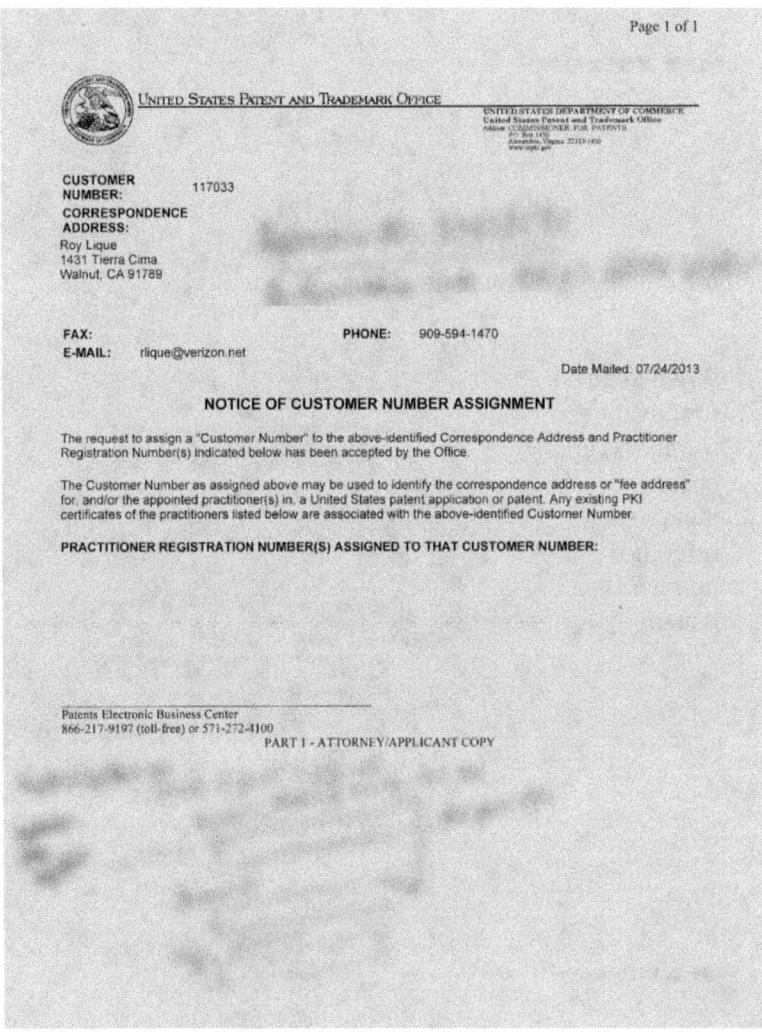

**Display Recovery Codes**

In case problems are encountered, 7 display recovery codes are available to restore system.

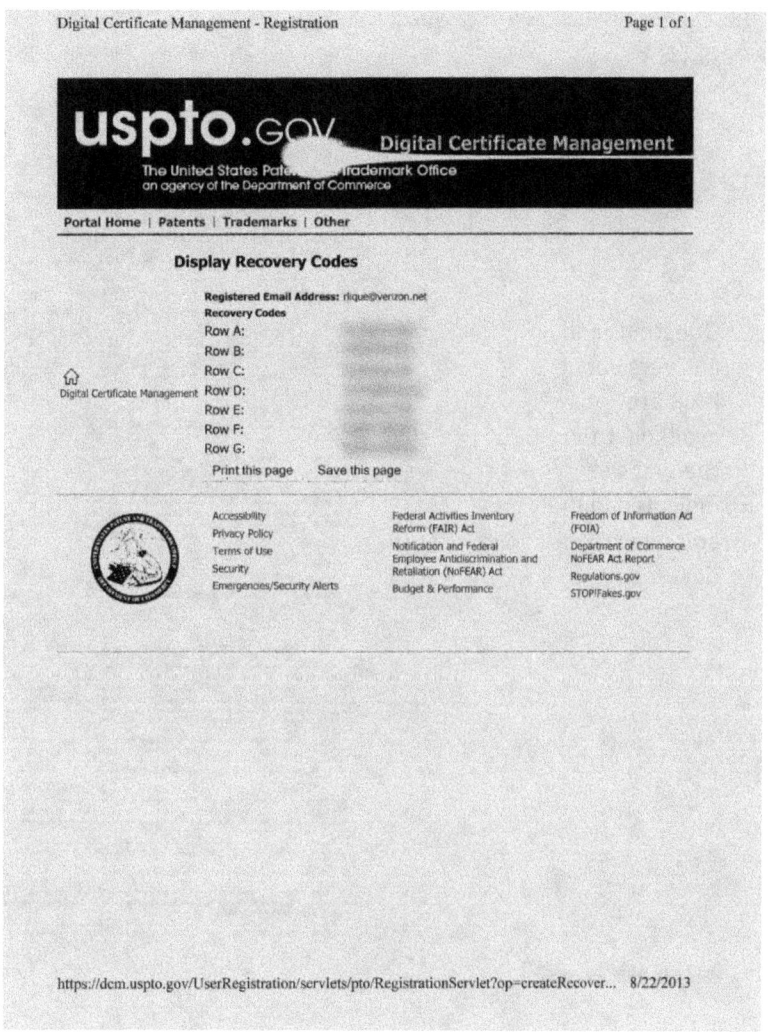

## Questions and Answers about PAIR

Questions and answers about PAIR are available from the website indicated in the adjoining document.

**20 of the Most Frequently Asked Questions from Customers**

**Questions Associated with becoming a Registered e-filing User**

**1.   How does a customer become a registered e-filer and gain access to Private PAIR?**

Becoming a registered e-filer is a two step process.   The first step is to request a customer number by filling out the Customer Number Request Form located at http://www.uspto.gov/web/forms/sb0125.pdf (http://www.uspto.gov/web/forms/sb0125.pdf). Either fax the Customer Number Request Form to the EBC at 571-273-0177 or return it by mail.   The EBC will process your customer number request within three business days.  After the customer number is received, the next step is to fill out, notarize, and mail in the Certificate Action Form. The Certificate Action form can be found at _____.   The PKI digital certificate is a small computer file that functions as a "digital identity" for a given user.   Once the EBC receives the Certificate Action Form, the Authorization Code will be emailed within five business days.  When the Authorization Code is received you may call the EBC directly and request the Reference Number.  These two codes together allow you to create your digital certificate file from our Digital Certificate Management site at _____.  This file will grant you access to Private PAIR and EFS-Web as a registered e-filer.

**2.   How can Applications be associated with a Customer Number?**

To associate existing patent applications to a Customer Number, you must complete and submit the Customer Number Upload Spreadsheet located at _____.   This information can also be sent on a CD or USB memory stick to the address listed below if many applications need to be associated.  Another method to associate an application to a customer number is to submit the Change of Correspondence Address form (SB/122) located at _____.   Unlike the Customer Number Upload Spreadsheet where multiple applications/patents can be listed, the Change of Correspondence Address form can be used to only associate a single application to a customer number.  Only the customer number should be indicated on the SB/122 form with physical address left blank.  Customer number associations using the SB/122 form are only processed by the Applications Assistants Unit or a Technology Center. The EBC processes the automated Upload Spreadsheet.

Mail Stop EBC Customer Number
Commissioner for Patents
P.O. Box 1450
Alexandria, VA 22313-1450

To associate a PCT application in the International phase with a Customer Number for purposes of viewing the PCT application in Private PAIR, please download and complete the _____ at _____ and mail or fax it to the EBC at 571-273-0177.

**3.   How can Customer Number data be corrected?**

Customer Number data may either be corrected electronically in Private PAIR or by submitting the USPTO 124A Customer Number Data Change form located at _____.   You may mail the form or fax it to 571-273-0177.  To use Private PAIR you must be a registered user.  Once Private PAIR is accessed go to the "Search by Customer Number" section and then select "View Customer Number Details."   Select the customer number from the drop down box and click search.  Select the "Request Customer Data Change(s)" button near the bottom of the web page to enter the Edit Customer Details screen.  Preview changes and transmit your request to the USPTO.

Adding or deleting a customer number that is associated with registration number will automatically add or delete the association between the PKI digital certificate assigned to the registration number and that customer number.

**4.   How can a Digital Certificate be created for new users?**

A Digital Certificate is created by the customer using two pieces of information from the USPTO.   One is an Authorization Code and the other is a Reference Number.  Each piece of information will be sent to you separately upon submitting a Certificate Action Form to the EBC.   Once you receive the Authorization Code you may call the EBC to obtain the Reference Number if you have not already received it. Once you have both the Authorization Code and Reference Number go to _____ and follow these steps:

1.  Click "New User", enter the Authorization Code and the Reference Number where requested.

2.  Click the "browse" button next to "digital certificate filename".

3.  When a dialog box appears, select the folder where you want the file to be saved at (somewhere you can locate the file easily, such as your desktop), then in the "filename" box, name your file (i.e. johndoe.epf) "making sure to leave the .EPF at the end of the file name, and then click "open".

4.  Enter the password that you wish to use twice following rules located in the box on the right side of the screen (when all rules are successfully followed green check marks will appear), then click "Submit".

http://www.uspto.gov/patents/ebc/top_questions_ebc.jsp                                   7/20/2013

## Sample Available Documents

All the documents pertaining to a patent application are available through PAIR.

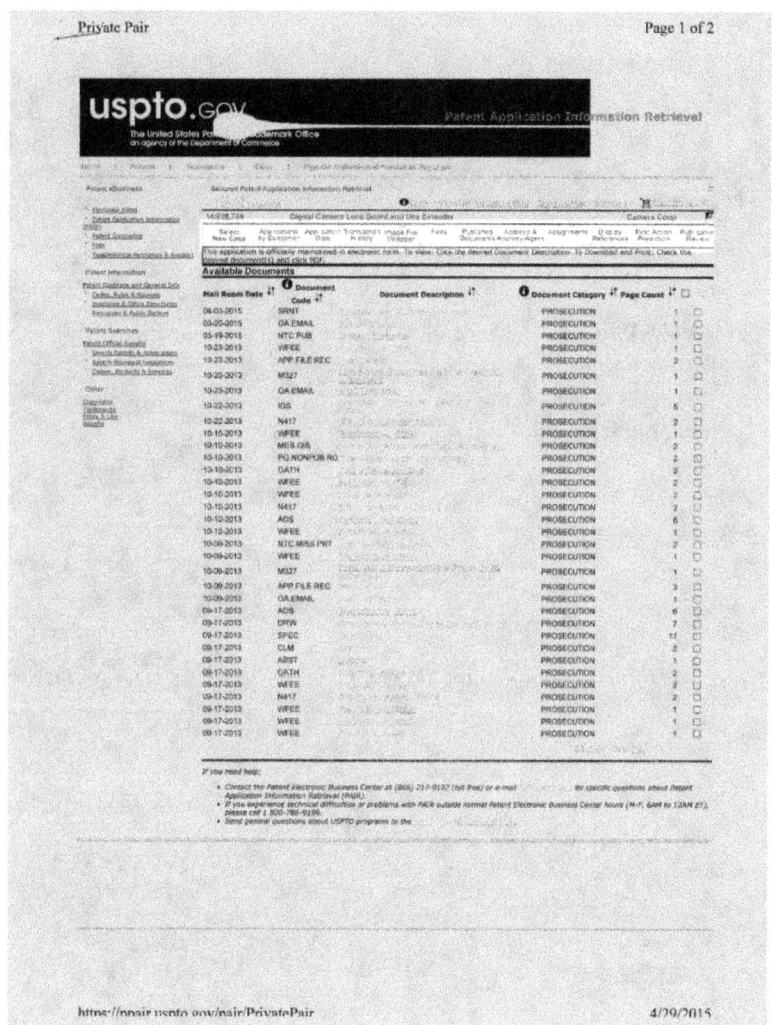

## Sample Transaction History

Transaction history is also readily available from PAIR.

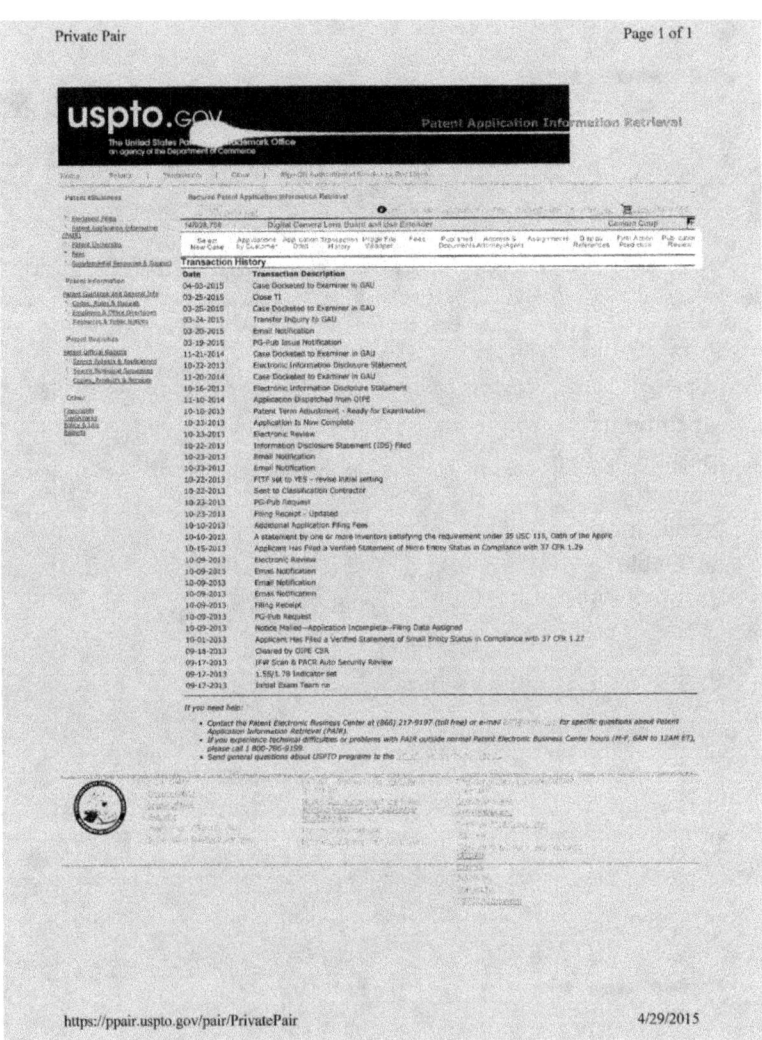

## Sample Payment History

Even the payment history pertaining to an application is readily available.

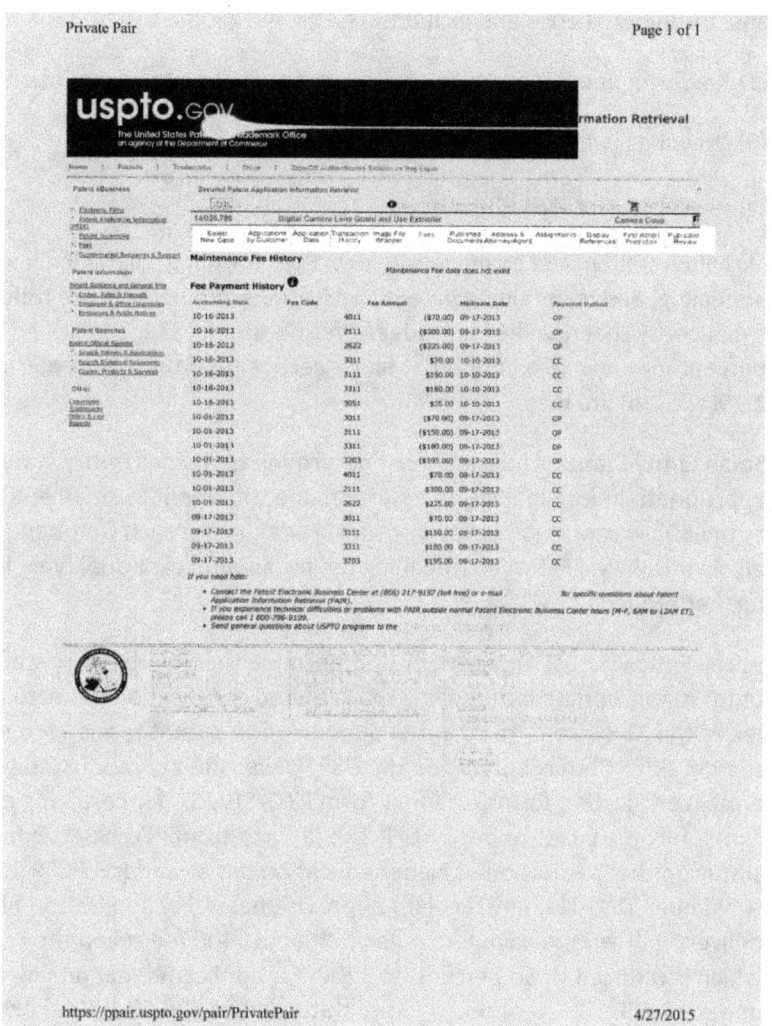

# Part 5 – Filing patent application – Non-Provisional

## Non Provisional Application

A nonprovisional application for a patent is made to the Director of the United States Patent and Trademark Office and includes:

(1) A written document which comprises a specification (description and claims);

(2) Drawings (when necessary);

(3) An oath or declaration; and

(4) Filing, search, and examination fees. Fees for filing, searching, examining, issuing, appealing, and maintaining patent applications and patents are reduced by 50 percent for any small entity that qualifies for reduced fees under 37 CFR 1.27(a), and are reduced by 75 percent for any micro entity that files a certification that the requirements under 37 CFR 1.29(a) or (d) are met.

Small Entity Status: Applicant must determine that small entity status under 37 CFR 1.27(a) is appropriate before making an assertion of entitlement to small entity status and paying a fee at the 50 percent small entity discount. Fees change each October. Note that by filing electronically via EFS-Web, the filing fee for an applicant qualifying for small entity status is further reduced.

Micro Entity Status: Applicant must determine that micro entity status under 37 CFR 1.29(a) or (d) is appropriate before filing the required certification of micro entity status and paying a fee at the 75 percent micro entity discount. The patent forms Web page is indexed under the section titled Forms, Patents on the USPTO website at www.uspto.gov. There are two micro entity certification forms – namely form PTO/SB/15A for certifying micro entity status on the "gross income basis" under 37 CFR 1.29(a), and form PTO/SB/15B for certifying micro entity status on the "institution of higher education basis" under 37 CFR 1.29(d). Effective November 15, 2011, any regular nonprovisional utility application filed by mail or hand-delivery will require payment of an additional $400 fee called the "non-electronic filing fee," which is reduced by 50 percent (to $200) for applicants that qualify for small entity status under 37 CFR 1.27(a) or micro entity status under 37 CFR 1.29(a) or (d). The only way to avoid having to pay the additional $400 non-electronic filing fee is by filing the regular nonprovisional utility application via EFS-Web.

Other patent correspondence, including design, plant, and provisional application filings, as well as correspondence filed in a nonprovisional application after the application filing date (known as "follow-on" correspondence), can still be filed by mail or hand-delivery without incurring the $400 non-electronic filing fee. You do not have to be a Registered eFiler to file a patent application via EFS-Web. However, unless you are a Registered eFiler, you must not attempt to file follow-on correspondence via EFS-Web, because Unregistered eFilers are not permitted to file follow-on correspondence via EFS-Web. Follow-on correspondence filed by anyone other than an EFS-Web Registered eFiler must be sent by mail or be hand-delivered. (See the "General Information and Correspondence" section of this brochure.) In the event you receive from the USPTO a "Notice of Incomplete Application" in response to your EFS-Web filing stating that an application number has been assigned but no filing date has been granted, you must become a Registered eFiler and file your reply to the "Notice of Incomplete Application" via EFS-Web in order to avoid the $400 non-electronic filing fee. To become a Registered eFiler and have the ability to file follow-on correspondence, please consult the information at www.uspto.gov/patents/process/file/efs/guidance/register.jsp, or call the Electronic Business Center at 866-217-9197.

The specification (description and claims) can be created using a word processing program such as Microsoft® Word or Corel® WordPerfect. The document containing the specification can normally be converted into PDF format by the word processing program itself so that it can be included as an attachment when filing the application via EFS-Web. Other application documents, such as drawings and a hand-signed declaration, may have to be scanned as a PDF file for filing via EFS-Web. See the information available at www.uspto.gov/patents/process/file/efs/index.jsp. Any questions regarding filing applications via EFS-Web should be directed to the Electronic Business Center at 866-217-9197.

All application documents must be in the English language or a translation into the English language will be required along with the required fee set forth in 37 CFR 1.17(i).

Each document (which should be filed via EFS-Web in PDF format) must have a top margin of at least 2 cm (3/4 inch), a left side margin of at least 2.5 cm (1 inch), a right side margin of at least 2 cm (3/4 inch) and a bottom margin of at least 2 cm (3/4 inch) with no holes made in the submitted papers. It is also required that the spacing on all papers be 1.5 or double-spaced and the application papers must be numbered consecutively (centrally located above or below the text) starting with page one.

The specification must have text written in a nonscript font (e.g., Arial, Times Roman, or Courier, preferably a font size of 12pt) lettering style having capital letters that should be at least 0.3175 cm (0.125 inch) high, but may be no smaller than 0.21 cm (0.08 inch) high (e.g., a font size of 6). The specification must have only a single column of text.

The specification must conclude with a claim or claims particularly pointing out and distinctly claiming the subject matter that the applicant regards as the invention. The portion of the application in which the applicant sets forth the claim or claims is an important part of the application, as it is the claims that define the scope of the protection afforded by the patent. The claims must commence on a separate sheet.

More than one claim may be presented provided they differ from each other. Claims may be presented in independent form (e.g. the claim stands by itself) or in dependent form, referring back to and further limiting another claim or claims in the same application. Any dependent claim that refers back to more than one other claim is considered a "multiple dependent claim."

The application for patent is not forwarded for examination until all required parts, complying with the rules related thereto, are received. If any application is filed without all the required parts for obtaining a filing date (incomplete or defective), the applicant will be notified of the deficiencies and given a time period to complete the application filing (a surcharge may be required)—at which time a filing date as of the date of such a completed submission will be obtained by the applicant. If the omission is not corrected within a specified time period, the application will be returned or otherwise disposed of; the filing fee if submitted will be refunded less a handling fee as set forth in the fee schedule.

The filing fee and declaration or oath need not be submitted with the parts requiring a filing date. It is, however, desirable that all parts of the complete application be deposited in the Office together; otherwise, each part must be signed and a letter must accompany each part, accurately and clearly connecting it with the other parts of the application. If an application that has been accorded a filing date does not include the filing fee or the oath or declaration, applicant will be notified and given a time period to pay the filing fee, file an oath or declaration and pay a surcharge.

All applications received in the USPTO are numbered in sequential order, and the applicant will be informed of the application number and filing date by a filing receipt.

The filing date of an application for patent is the date on which a specification (including at least one claim) and any drawings necessary to understand the subject matter sought to be patented are received in the USPTO; or the date on which the last part completing the application is received in the case of a previously incomplete or defective application.

## Arrangement of application elements

- (a) The elements of the application, if applicable, should appear in the following order:
  - (1) Utility application transmittal form.
  - (2) Fee transmittal form.
  - (3) Application data sheet (see § **1.76** ).
  - (4) Specification.
  - (5) Drawings.
  - (6) The inventor's oath or declaration.

- (b) The specification should include the following sections in order:
  - (1) Title of the invention, which may be accompanied by an introductory portion stating the name, citizenship, and residence of the applicant (unless included in the application data sheet).
  - (2) Cross-reference to related applications.
  - (3) Statement regarding federally sponsored research or development.
  - (4) The names of the parties to a joint research agreement.
  - (5) Reference to a "Sequence Listing," a table, or a computer program listing appendix submitted on a compact disc and an incorporation-by-reference of the material on the compact disc (see § **1.52(e)(5)**). The total number of compact discs including duplicates and the files on each compact disc shall be specified.
  - (6) Statement regarding prior disclosures by the inventor or a joint inventor.
  - (7) Background of the invention.
  - (8) Brief summary of the invention.
  - (9) Brief description of the several views of the drawing.
  - (10) Detailed description of the invention.
  - (11) A claim or claims.
  - (12) Abstract of the disclosure.
  - (13) "Sequence Listing," if on paper (see §§ **1.821** through **1.825** ).
- (c) The text of the specification sections defined in paragraphs (b)(1) through (b)(12) of this section, if applicable, should be preceded by a section heading in uppercase and without underlining or bold type.

## Electronic Acknowledgement

## Electronic Acknowledgement Receipt – Page 1

| Electronic Acknowledgement Receipt | |
|---|---|
| EFS ID: | 16873732 |
| Application Number: | 14028786 |
| International Application Number: | |
| Confirmation Number: | 6716 |
| Title of Invention: | Digital Camera Lens Guard and Use Extender |
| First Named Inventor/Applicant Name: | Roy Lique |
| Customer Number: | 117033 |
| Filer: | Roy Lique |
| Filer Authorized By: | |
| Attorney Docket Number: | Camera Coup |
| Receipt Date: | 17-SEP-2013 |
| Filing Date: | |
| Time Stamp: | 13:25:25 |
| Application Type: | Utility under 35 USC 111(a) |

**Payment information:**

| | |
|---|---|
| Submitted with Payment | yes |
| Payment Type | Credit Card |
| Payment was successfully received in RAM | $595 |
| RAM confirmation Number | 12522 |
| Deposit Account | |
| Authorized User | |

**File Listing:**

| Document Number | Document Description | File Name | File Size(Bytes)/ Message Digest | Multi Part /.zip | Pages (if appl.) |
|---|---|---|---|---|---|

## Electronic Acknowledgement Receipt – Page 2

| 1 | Application Data Sheet | ApplDataSheetOfficeOrg.pdf | 1504695 <br> a104f00a7aeddd4860234b56b01b7fac4488a | no | 6 |
|---|---|---|---|---|---|
| **Warnings:** | | | | | |
| **Information:** | | | | | |
| 2 | Drawings-only black and white line drawings | DrawingOfficeOrg.pdf | 397871 <br> 1ce7d8e4c1ad5d9b1f1d566334ecd61cb9f4889b | no | 7 |
| **Warnings:** | | | | | |
| **Information:** | | | | | |
| 3 | Specification | SpecificationOfficeOrg.pdf | 108267 <br> cfc90cf309c4f2b0da0b631f5d95cdbcdeaf15f | no | 20 |
| **Warnings:** | | | | | |
| **Information:** | | | | | |
| 4 | Oath or Declaration filed | OathDeclarationOfficeOrg.pdf | 75713 <br> d1966e5c9915a6a4mb1Zb41a6b05b03af9bf | no | 2 |
| **Warnings:** | | | | | |
| **Information:** | | | | | |
| 5 | Fee Worksheet (SB06) | fee-info.pdf | 36334 <br> a934d298b5d0c43c3d2d1c374b019109b2d0726 | no | 2 |
| **Warnings:** | | | | | |
| **Information:** | | | | | |
| | | **Total Files Size (in bytes):** | | 2122880 | |

This Acknowledgement Receipt evidences receipt on the noted date by the USPTO of the indicated documents, characterized by the applicant, and including page counts, where applicable. It serves as evidence of receipt similar to a Post Card, as described in MPEP 503.

**New Applications Under 35 U.S.C. 111**
If a new application is being filed and the application includes the necessary components for a filing date (see 37 CFR 1.53(b)-(d) and MPEP 506), a Filing Receipt (37 CFR 1.54) will be issued in due course and the date shown on this Acknowledgement Receipt will establish the filing date of the application.

**National Stage of an International Application under 35 U.S.C. 371**
If a timely submission to enter the national stage of an international application is compliant with the conditions of 35 U.S.C. 371 and other applicable requirements a Form PCT/DO/EO/903 indicating acceptance of the application as a national stage submission under 35 U.S.C. 371 will be issued in addition to the Filing Receipt, in due course.

**New International Application Filed with the USPTO as a Receiving Office**
If a new international application is being filed and the international application includes the necessary components for an international filing date (see PCT Article 11 and MPEP 1810), a Notification of the International Application Number and of the International Filing Date (Form PCT/RO/105) will be issued in due course, subject to prescriptions concerning national security, and the date shown on this Acknowledgement Receipt will establish the international filing date of the application.

## Electronic Acknowledgement Receipt – Page 3

| Total Files Size (in bytes): | 130738 |
|---|---|

This Acknowledgement Receipt evidences receipt on the noted date by the USPTO of the indicated documents, characterized by the applicant, and including page counts, where applicable. It serves as evidence of receipt similar to a Post Card, as described in MPEP 503.

New Applications Under 35 U.S.C. 111
If a new application is being filed and the application includes the necessary components for a filing date (see 37 CFR 1.53(b)-(d) and MPEP 506), a Filing Receipt (37 CFR 1.54) will be issued in due course and the date shown on this Acknowledgement Receipt will establish the filing date of the application.

National Stage of an International Application under 35 U.S.C. 371
If a timely submission to enter the national stage of an international application is compliant with the conditions of 35 U.S.C. 371 and other applicable requirements a Form PCT/DO/EO/903 indicating acceptance of the application as a national stage submission under 35 U.S.C. 371 will be issued in addition to the Filing Receipt, in due course.

New International Application Filed with the USPTO as a Receiving Office
If a new international application is being filed and the international application includes the necessary components for an international filing date (see PCT Article 11 and MPEP 1810), a Notification of the International Application Number and of the International Filing Date (Form PCT/RO/105) will be issued in due course, subject to prescriptions concerning national security, and the date shown on this Acknowledgement Receipt will establish the international filing date of the application.

# Fee Determination

## Fee Determination – Page 1

Doc Code: WFEE
Document Description: Fee Worksheet (PTO-875)

PTO/SB/06 (03-13)
Approved for use through 01/31/2014. OMB 0651-0032
U.S. Patent and Trademark Office, U.S. DEPARTMENT OF COMMERCE

Under the Paperwork Reduction Act of 1995, no persons are required to respond to a collection of information unless it displays a valid OMB control number.

**PATENT APPLICATION FEE DETERMINATION RECORD**
Substitute for Form PTO-875

Application or Docket Number: 14/028,786

### APPLICATION AS FILED – PART I

| FOR | NUMBER FILED | NUMBER EXTRA | RATE ($) | FEE ($) |
|---|---|---|---|---|
| BASIC FEE | N/A | N/A | N/A | 70.00 |
| SEARCH FEE | N/A | N/A | N/A | 150.00 |
| EXAMINATION FEE | N/A | N/A | N/A | 180.00 |
| TOTAL CLAIMS | 10 minus 20 = | *0 | x | 0.00 |
| INDEPENDENT CLAIMS | minus 3 = | *0 | x | 0.00 |
| APPLICATION SIZE FEE | | | | 0.00 |
| MULTIPLE DEPENDENT CLAIM PRESENT | | | N/A | 195.00 |
| | | | TOTAL | 595.00 |

MICRO ENTITY selected.

117

# Fee Determination – Page 2

## Privacy Act Statement

**The Privacy Act of 1974 (P.L. 93-579)** requires that you be given certain information in connection with your submission of the attached form related to a patent application or patent. Accordingly, pursuant to the requirements of the Act, please be advised that: (1) the general authority for the collection of this information is 35 U.S.C. 2(b)(2); (2) furnishing of the information solicited is voluntary; and (3) the principal purpose for which the information is used by the U.S. Patent and Trademark Office is to process and/or examine your submission related to a patent application or patent. If you do not furnish the requested information, the U.S. Patent and Trademark Office may not be able to process and/or examine your submission, which may result in termination of proceedings or abandonment of the application or expiration of the patent.

The information provided by you in this form will be subject to the following routine uses:

1. The information on this form will be treated confidentially to the extent allowed under the Freedom of Information Act (5 U.S.C. 552) and the Privacy Act (5 U.S.C 552a). Records from this system of records may be disclosed to the Department of Justice to determine whether disclosure of these records is required by the Freedom of Information Act.
2. A record from this system of records may be disclosed, as a routine use, in the course of presenting evidence to a court, magistrate, or administrative tribunal, including disclosures to opposing counsel in the course of settlement negotiations.
3. A record in this system of records may be disclosed, as a routine use, to a Member of Congress submitting a request involving an individual, to whom the record pertains, when the individual has requested assistance from the Member with respect to the subject matter of the record.
4. A record in this system of records may be disclosed, as a routine use, to a contractor of the Agency having need for the information in order to perform a contract. Recipients of information shall be required to comply with the requirements of the Privacy Act of 1974, as amended, pursuant to 5 U.S.C. 552a(m).
5. A record related to an International Application filed under the Patent Cooperation Treaty in this system of records may be disclosed, as a routine use, to the International Bureau of the World Intellectual Property Organization, pursuant to the Patent Cooperation Treaty.
6. A record in this system of records may be disclosed, as a routine use, to another federal agency for purposes of National Security review (35 U.S.C. 181) and for review pursuant to the Atomic Energy Act (42 U.S.C. 218(c)).
7. A record from this system of records may be disclosed, as a routine use, to the Administrator, General Services, or his/her designee, during an inspection of records conducted by GSA as part of that agency's responsibility to recommend improvements in records management practices and programs, under authority of 44 U.S.C. 2904 and 2906. Such disclosure shall be made in accordance with the GSA regulations governing inspection of records for this purpose, and any other relevant (i.e., GSA or Commerce) directive. Such disclosure shall not be used to make determinations about individuals.
8. A record from this system of records may be disclosed, as a routine use, to the public after either publication of the application pursuant to 35 U.S.C. 122(b) or issuance of a patent pursuant to 35 U.S.C. 151. Further, a record may be disclosed, subject to the limitations of 37 CFR 1.14, as a routine use, to the public if the record was filed in an application which became abandoned or in which the proceedings were terminated and which application is referenced by either a published application, an application open to public inspection or an issued patent.
9. A record from this system of records may be disclosed, as a routine use, to a Federal, State, or local law enforcement agency, if the USPTO becomes aware of a violation or potential violation of law or regulation.

## Fee Transmittal

## Fee Transmittal – Page 1

Degree of difficulty of filling highlighted items: EASY

| Electronic Patent Application Fee Transmittal | |
|---|---|
| Application Number: | 14028786 |
| Filing Date: | 17-Sep-2013 |
| Title of Invention: | Digital Camera Lens Guard and Use Extender |
| First Named Inventor/Applicant Name: | Roy Lique |
| Filer: | Roy Lique |
| Attorney Docket Number: | Camera Coup |

Filed as Micro Entity

**Utility under 35 USC 111(a) Filing Fees**

| Description | Fee Code | Quantity | Amount | Sub-Total in USD($) |
|---|---|---|---|---|
| **Basic Filing:** | | | | |
| BASIC UTILITY PATENT FILING FEE-MICRO-ENT | 3011 | 1 | 70 | 70 |
| Utility Search Fee | 3111 | 1 | 150 | 150 |
| Utility Examination Fee | 3311 | 1 | 180 | 180 |
| **Pages:** | | | | |
| **Claims:** | | | | |
| Multiple Dependent Claims | 3203 | 1 | 195 | 195 |
| **Miscellaneous-Filing:** | | | | |
| **Petition:** | | | | |

119

## Fee Transmittal – Page 2

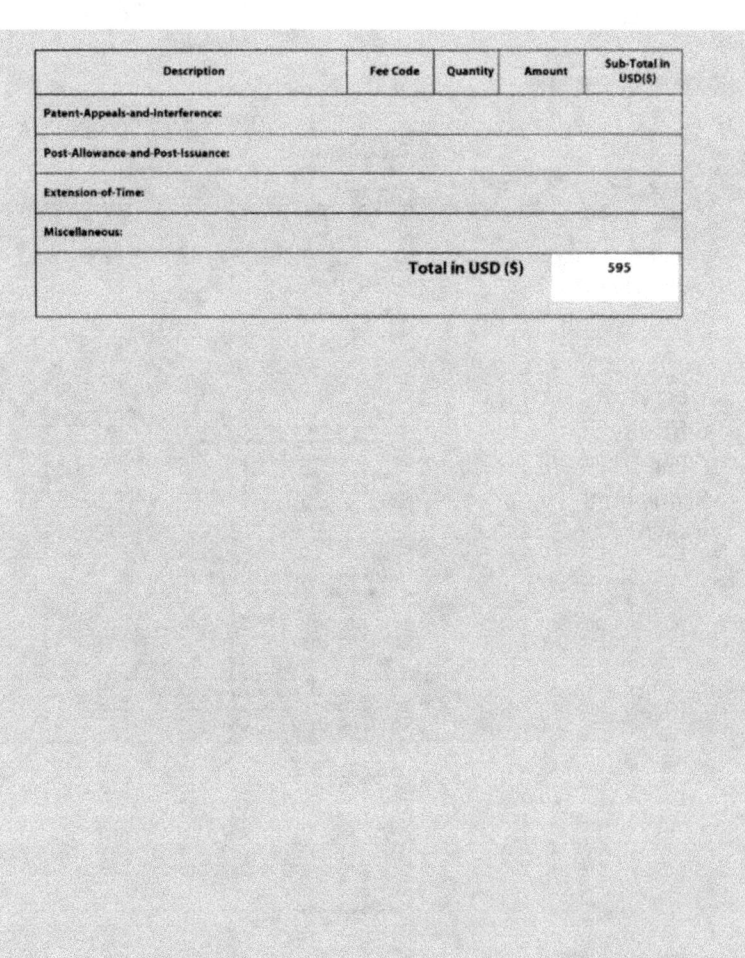

| Description | Fee Code | Quantity | Amount | Sub-Total in USD($) |
|---|---|---|---|---|
| Patent-Appeals-and-Interference: | | | | |
| Post-Allowance-and-Post-Issuance: | | | | |
| Extension-of-Time: | | | | |
| Miscellaneous: | | | | |
| Total in USD ($) | | | | 595 |

Degree of difficulty of filling highlighted items: EASY

# Application Data Sheet

## Application Data Sheet – Page 1

Degree of difficulty of filling highlighted items: EASY

PTO/AIA/14 (03-13)
Approved for use through 01/31/2014. OMB 0651-0032
U.S. Patent and Trademark Office; U.S. DEPARTMENT OF COMMERCE
Under the Paperwork Reduction Act of 1995, no persons are required to respond to a collection of information unless it contains a valid OMB control number.

| Application Data Sheet 37 CFR 1.76 | Attorney Docket Number | Camera Coup |
| | Application Number | |

| Title of Invention | Digital Camera Lens Guard and Use Extender |

The application data sheet is part of the provisional or nonprovisional application for which it is being submitted. The following form contains the bibliographic data arranged in a format specified by the United States Patent and Trademark Office as outlined in 37 CFR 1.76. This document may be completed electronically and submitted to the Office in electronic format using the Electronic Filing System (EFS) or the document may be printed and included in a paper filed application.

## Secrecy Order 37 CFR 5.2

☐ Portions or all of the application associated with this Application Data Sheet may fall under a Secrecy Order pursuant to 37 CFR 5.2 (Paper filers only. Applications that fall under Secrecy Order may not be filed electronically.)

## Inventor Information:

| Inventor 1 | | | | | Remove |
| Legal Name | | | | | |

| Prefix | Given Name | Middle Name | Family Name | Suffix |
|---|---|---|---|---|
| Mr. | Roy | | Lique | |

Residence Information (Select One)  ◉ US Residency  ○ Non US Residency  ○ Active US Military Service

| City | Walnut | State/Province | CA | Country of Residence | US |

**Mailing Address of Inventor:**

| Address 1 | 1431 Tierra Cima | | |
| Address 2 | | | |
| City | Walnut | State/Province | CA |
| Postal Code | 91789 | Country i | US |

All Inventors Must Be Listed - Additional Inventor Information blocks may be generated within this form by selecting the **Add** button. [Add]

## Correspondence Information:

Enter either Customer Number or complete the Correspondence Information section below. For further information see 37 CFR 1.33(a).

☐ An Address is being provided for the correspondence Information of this application.

| Customer Number | 117033 | | |
| Email Address | rlique@verizon.net | [Add Email] | [Remove Email] |

## Application Information:

| Title of the Invention | Digital Camera Lens Guard and Use Extender | | |
| Attorney Docket Number | Camera Coup | Small Entity Status Claimed | ☒ |
| Application Type | Nonprovisional | | |
| Subject Matter | Utility | | |
| Total Number of Drawing Sheets (if any) | 7 | Suggested Figure for Publication (if any) | 1 |

EFS Web 2.2.8

## Application Data Sheet – Page 2

**Degree of difficulty of filling highlighted items: EASY**

---

PTO/AIA/14 (03-13)
Approved for use through 01/31/2014. OMB 0651-0032
U.S. Patent and Trademark Office; U.S. DEPARTMENT OF COMMERCE
Under the Paperwork Reduction Act of 1995, no persons are required to respond to a collection of information unless it contains a valid OMB control number.

| Application Data Sheet 37 CFR 1.76 | Attorney Docket Number | Camera Coup |
|---|---|---|
| | Application Number | |

| Title of Invention | Digital Camera Lens Guard and Use Extender |
|---|---|

**Publication Information:**

☐ Request Early Publication (Fee required at time of Request 37 CFR 1.219)

☒ **Request Not to Publish.** I hereby request that the attached application not be published under 35 U.S.C. 122(b) and certify that the invention disclosed in the attached application **has not and will not** be the subject of an application filed in another country, or under a multilateral international agreement, that requires publication at eighteen months after filing.

**Representative Information:**

Representative information should be provided for all practitioners having a power of attorney in the application. Providing this information in the Application Data Sheet does not constitute a power of attorney in the application (see 37 CFR 1.32). Either enter Customer Number or complete the Representative Name section below. If both sections are completed the customer Number will be used for the Representative Information during processing.

| Please Select One: | ⦿ Customer Number | ◯ US Patent Practitioner | ◯ Limited Recognition (37 CFR 11.9) |
|---|---|---|---|
| Customer Number | | | |

**Domestic Benefit/National Stage Information:**

This section allows for the applicant to either claim benefit under 35 U.S.C. 119(e), 120, 121, or 365(c) or indicate National Stage entry from a PCT application. Providing this information in the application data sheet constitutes the specific reference required by 35 U.S.C. 119(e) or 120, and 37 CFR 1.78.

| Prior Application Status | | | Remove |
|---|---|---|---|
| Application Number | Continuity Type | Prior Application Number | Filing Date (YYYY-MM-DD) |

Additional Domestic Benefit/National Stage Data may be generated within this form by selecting the **Add** button.

**Foreign Priority Information:**

This section allows for the applicant to claim priority to a foreign application. Providing this information in the application data sheet constitutes the claim for priority as required by 35 U.S.C. 119(b) and 37 CFR 1.55(d). When priority is claimed to a foreign application that is eligible for retrieval under the priority document exchange program (PDX) the information will be used by the Office to automatically attempt retrieval pursuant to 37 CFR 1.55(h)(1) and (2). Under the PDX program, applicant bears the ultimate responsibility for ensuring that a copy of the foreign application is received by the Office from the participating foreign intellectual property office, or a certified copy of the foreign priority application is filed, within the time period specified in 37 CFR 1.55(g)(1).

| | | | Remove |
|---|---|---|---|
| Application Number | Country | Filing Date (YYYY-MM-DD) | Access Code (if applicable) |

Additional Foreign Priority Data may be generated within this form by selecting the **Add** button.

EFS Web 2.2.8

## Application Data Sheet – Page 3

Degree of difficulty of filling highlighted items: EASY

PTO/AIA/14 (03-13)
Approved for use through 01/31/2014. OMB 0661-0032
U.S. Patent and Trademark Office; U.S. DEPARTMENT OF COMMERCE
Under the Paperwork Reduction Act of 1995, no persons are required to respond to a collection of information unless it contains a valid OMB control number.

| Application Data Sheet 37 CFR 1.76 | Attorney Docket Number | Camera Coup |
|---|---|---|
| | Application Number | |

| Title of Invention | Digital Camera Lens Guard and Use Extender |
|---|---|

### Statement under 37 CFR 1.55 or 1.78 for AIA (First Inventor to File) Transition Applications

☒ This application (1) claims priority to or the benefit of an application filed before March 16, 2013 and (2) also contains, or contained at any time, a claim to a claimed invention that has an effective filing date on or after March 16, 2013.
NOTE: By providing this statement under 37 CFR 1.55 or 1.78, this application, with a filing date on or after March 16, 2013, will be examined under the first inventor to file provisions of the AIA.

### Authorization to Permit Access:

☐ Authorization to Permit Access to the Instant Application by the Participating Offices

If checked, the undersigned hereby grants the USPTO authority to provide the European Patent Office (EPO), the Japan Patent Office (JPO), the Korean Intellectual Property Office (KIPO), the World Intellectual Property Office (WIPO), and any other intellectual property offices in which a foreign application claiming priority to the instant patent application is filed access to the instant patent application. See 37 CFR 1.14(c) and (h). This box should not be checked if the applicant does not wish the EPO, JPO, KIPO, WIPO, or other intellectual property office in which a foreign application claiming priority to the instant patent application is filed to have access to the instant patent application.

In accordance with 37 CFR 1.14(h)(3), access will be provided to a copy of the instant patent application with respect to: 1) the instant patent application-as-filed, 2) any foreign application to which the instant patent application claims priority under 35 U.S.C. 119(a)-(d) if a copy of the foreign application that satisfies the certified copy requirement of 37 CFR 1.55 has been filed in the instant patent application; and 3) any U.S. application-as-filed from which benefit is sought in the instant patent application.

In accordance with 37 CFR 1.14(c), access may be provided to information concerning the date of filing this Authorization.

### Applicant Information:

Providing assignment information in this section does not substitute for compliance with any requirement of part 3 of Title 37 of CFR to have an assignment recorded by the Office.

EFS Web 2.2.9

## Application Data Sheet – Page 4

Degree of difficulty of filling highlighted items: EASY

PTO/AIA/14 (03-13)
Approved for use through 01/31/2014. OMB 0651-0032
U.S. Patent and Trademark Office; U.S. DEPARTMENT OF COMMERCE
Under the Paperwork Reduction Act of 1995, no persons are required to respond to a collection of information unless it contains a valid OMB control number.

| Application Data Sheet 37 CFR 1.76 | Attorney Docket Number | Camera Coup |
| | Application Number | |

| Title of Invention | Digital Camera Lens Guard and Use Extender |

**Applicant   1**

If the applicant is the inventor (or the remaining joint inventor or inventors under 37 CFR 1.45), this section should not be completed. The information to be provided in this section is the name and address of the legal representative who is the applicant under 37 CFR 1.43, or the name and address of the assignee, person to whom the inventor is under an obligation to assign the invention, or person who otherwise shows sufficient proprietary interest in the matter who is the applicant under 37 CFR 1.46. If the applicant is an applicant under 37 CFR 1.46 (assignee, person to whom the inventor is obligated to assign, or person who otherwise shows sufficient proprietary interest) together with one or more joint inventors, then the joint inventor or inventors who are also the applicant should be identified in this section.

Clear

| ○ Assignee | ○ Legal Representative under 35 U.S.C. 117 | ○ Joint Inventor |

| ○ Person to whom the inventor is obligated to assign. | ○ Person who shows sufficient proprietary interest |

If applicant is the legal representative, indicate the authority to file the patent application, the inventor is:

Name of the Deceased or Legally Incapacitated Inventor :

If the Applicant is an Organization check here.   ☐

| Prefix | Given Name | Middle Name | Family Name | Suffix |
| | | | | |

**Mailing Address Information For Applicant:**

| Address 1 | |
| Address 2 | |
| City | | State/Province | |
| Country | | Postal Code | |
| Phone Number | | Fax Number | |
| Email Address | |

Additional Applicant Data may be generated within this form by selecting the Add button.

## Assignee Information including Non-Applicant Assignee Information:

Providing assignment information in this section does not substitute for compliance with any requirement of part 3 of Title 37 of CFR to have an assignment recorded by the Office.

**Assignee   1**

Complete this section if assignee information, including non-applicant assignee information, is desired to be included on the patent application publication . An assignee-applicant identified in the "Applicant Information" section will appear on the patent application publication as an applicant. For an assignee-applicant, complete this section only if identification as an assignee is also desired on the patent application publication.

EFS Web 2.2.8

## Application Data Sheet – Page 5

Degree of difficulty of filling highlighted items: EASY

Signature name must be immediately preceded by a forward slash and immediately followed by another forward slash

PTO/AIA/14 (03-13)
Approved for use through 01/31/2014. OMB 0651-0032
U.S. Patent and Trademark Office; U.S. DEPARTMENT OF COMMERCE
Under the Paperwork Reduction Act of 1995, no persons are required to respond to a collection of information unless it contains a valid OMB control number.

| Application Data Sheet 37 CFR 1.76 | Attorney Docket Number | Camera Coup |
| | Application Number | |

| Title of Invention | Digital Camera Lens Guard and Use Extender |

If the Assignee is an Organization check here. ☐

| Prefix | Given Name | Middle Name | Family Name | Suffix |
|---|---|---|---|---|
| | | | | |

**Mailing Address Information For Non-Applicant Assignee:**

| Address 1 | |
| Address 2 | |
| City | | State/Province | |
| Country | | Postal Code | |
| Phone Number | | Fax Number | |
| Email Address | |

Additional Assignee Data may be generated within this form by selecting the Add button. [ Add ]

**Signature:** [ Remove ]

NOTE: This form must be signed in accordance with 37 CFR 1.33. See 37 CFR 1.4 for signature requirements and certifications.

| Signature | /Roy Lique/ | | | Date (YYYY-MM-DD) | 2013-09-15 |
|---|---|---|---|---|---|
| First Name | Roy | Last Name | Lique | Registration Number | |

Additional Signature may be generated within this form by selecting the Add button. [ Add ]

This collection of information is required by 37 CFR 1.76. The information is required to obtain or retain a benefit by the public which is to file (and by the USPTO to process) an application. Confidentiality is governed by 35 U.S.C. 122 and 37 CFR 1.14. This collection is estimated to take 23 minutes to complete, including gathering, preparing, and submitting the completed application data sheet form to the USPTO. Time will vary depending upon the individual case. Any comments on the amount of time you require to complete this form and/or suggestions for reducing this burden, should be sent to the Chief Information Officer, U.S. Patent and Trademark Office, U.S. Department of Commerce, P.O. Box 1450, Alexandria, VA 22313-1450. DO NOT SEND FEES OR COMPLETED FORMS TO THIS ADDRESS. SEND TO: Commissioner for Patents, P.O. Box 1450, Alexandria, VA 22313-1450.

EFS Web 2.2.8

## Application Data Sheet – Page 6

**Privacy Statement**

**Privacy Act Statement**

The Privacy Act of 1974 (P.L. 93-579) requires that you be given certain information in connection with your submission of the attached form related to a patent application or patent. Accordingly, pursuant to the requirements of the Act, please be advised that: (1) the general authority for the collection of this information is 35 U.S.C. 2(b)(2); (2) furnishing of the information solicited is voluntary; and (3) the principal purpose for which the information is used by the U.S. Patent and Trademark Office is to process and/or examine your submission related to a patent application or patent. If you do not furnish the requested information, the U.S. Patent and Trademark Office may not be able to process and/or examine your submission, which may result in termination of proceedings or abandonment of the application or expiration of the patent.

The information provided by you in this form will be subject to the following routine uses:

1. The information on this form will be treated confidentially to the extent allowed under the Freedom of Information Act (5 U.S.C. 552) and the Privacy Act (5 U.S.C. 552a). Records from this system of records may be disclosed to the Department of Justice to determine whether the Freedom of Information Act requires disclosure of these records.

2. A record from this system of records may be disclosed, as a routine use, in the course of presenting evidence to a court, magistrate, or administrative tribunal, including disclosures to opposing counsel in the course of settlement negotiations.

3. A record in this system of records may be disclosed, as a routine use, to a Member of Congress submitting a request involving an individual, to whom the record pertains, when the individual has requested assistance from the Member with respect to the subject matter of the record.

4. A record in this system of records may be disclosed, as a routine use, to a contractor of the Agency having need for the information in order to perform a contract. Recipients of information shall be required to comply with the requirements of the Privacy Act of 1974, as amended, pursuant to 5 U.S.C. 552a(m).

5. A record related to an International Application filed under the Patent Cooperation Treaty in this system of records may be disclosed, as a routine use, to the International Bureau of the World Intellectual Property Organization, pursuant to the Patent Cooperation Treaty.

6. A record in this system of records may be disclosed, as a routine use, to another federal agency for purposes of National Security review (35 U.S.C. 181) and for review pursuant to the Atomic Energy Act (42 U.S.C. 218(c)).

7. A record from this system of records may be disclosed, as a routine use, to the Administrator, General Services, or his/her designee, during an inspection of records conducted by GSA as part of that agency's responsibility to recommend improvements in records management practices and programs, under authority of 44 U.S.C. 2904 and 2906. Such disclosure shall be made in accordance with the GSA regulations governing inspection of records for this purpose, and any other relevant (i.e., GSA or Commerce) directive. Such disclosure shall not be used to make determinations about individuals.

8. A record from this system of records may be disclosed, as a routine use, to the public after either publication of the application pursuant to 35 U.S.C. 122(b) or issuance of a patent pursuant to 35 U.S.C. 151. Further, a record may be disclosed, subject to the limitations of 37 CFR 1.14, as a routine use, to the public if the record was filed in an application which became abandoned or in which the proceedings were terminated and which application is referenced by either a published application, an application open to public inspections or an issued patent.

9. A record from this system of records may be disclosed, as a routine use, to a Federal, State, or local law enforcement agency, if the USPTO becomes aware of a violation or potential violation of law or regulation.

EFS Web 2.2.8

126

## Non-Provisional Application - Specifications – Provisions that apply

The specification should have the following sections, in order:

(1) Title of the Invention

(7) Brief Summary of the Invention

(8) Brief description of the several views of the drawing (if any)

(9) Detailed Description of the Invention

(10) A claim or claims

(11) Abstract of the disclosure

The specification must include a written description of the invention and of the manner and process of making and using it, and is required to be in such full, clear, concise, and exact terms as to enable any person skilled in the technological area to which the invention pertains, or with which it is most nearly connected, to make and use the same.

The specification must set forth the precise invention for which a patent is solicited, in such manner as to distinguish it from other inventions and from what is old. It must describe completely a specific embodiment of the process, machine, manufacture, composition of matter, or improvement invented, and must explain the mode of operation or principle whenever applicable. The best mode contemplated by the inventor for carrying out the invention must be set forth.

The title of the invention, which should be as short and specific as possible (no more than 500 characters), should appear as a heading on the first page of the specification if it does not otherwise appear at the beginning of the application. A brief abstract of the technical disclosure in the specification, including that which is new in the art to which the invention pertains, must be set forth on a separate page preferably following the claims. The abstract should be in the form of a single paragraph of 150 words or less, under the heading "Abstract of the Disclosure."

A brief summary of the invention indicating its nature and substance, which may include a statement of the object of the invention, should precede the detailed description. The summary should be commensurate with the invention as claimed, and any object recited should be that of the invention as claimed.

When there are drawings, there shall be a brief description of the several views of the drawings, and the detailed description of the invention shall refer to the different views by specifying the numbers of the figures, and to the different parts by use of reference numerals.

The specification must conclude with a claim or claims particularly pointing out and distinctly claiming the subject matter that the applicant regards as the invention. The portion of the application in which the applicant sets forth the claim or claims is an important part of the application, as it is the claims that define the scope of the protection afforded by the patent and which questions of infringement are judged by the courts.

More than one claim may be presented, provided they differ substantially from each other and are not unduly multiplied. One or more claims may be presented in dependent form, referring back to and further limiting another claim or claims in the same application. Any dependent claim that refers back to more than one other claim is considered a "multiple dependent claim."

Multiple dependent claims shall refer to such other claims in the alternative only. A multiple dependent claim shall not serve as a basis for any other multiple dependent claims. Claims in dependent form shall be construed to include all of the limitations of the claim incorporated by reference into the dependent claim. A multiple dependent claim shall be construed to incorporate all the limitations of each of the particular claims in relation to which it is being considered.

The claim or claims must conform to the invention as set forth in the remainder of the specification and the terms and phrases used in the claims must find clear support or antecedent basis in the description so that the meaning of the terms in the claims may be ascertainable by reference to the description.

## Non-provisional Application - Specification - How I developed it

I could write an equally effective specification. That was the conclusion I derived from reviewing a few existing ones for different inventions, to find out if I was ready for the task. I did not sit down and start writing without having an idea how others did with their applications for patents. It helped me a great deal to download a few of them and make them available for comparison.

Putting the project at hand in perspective, I had never written any type of specifications before. Since English is not my native language, the task appeared to be even more daunting. But by diligently following the sample specification in "Patent It Yourself", from Nolo Press, by Attorney David Pressman, my resolve to do the job was absolute.

As with the other parts of this book, the specification accompanying the application to secure the patent for the Digital Camera Lens Guard and Use Extender was modeled after the sample one in the previously mentioned reference book. I took it upon myself to make some adjustments in my written material to conform to the USPTO requirements.

In writing the specification, I stuck with the PHOSITA (person having ordinary skill in the art) abbreviation and because of it, I was seldom sidetracked. With PHOSITA, I was able to define my audience and write accordingly. When, for example, I tended to blindly use some legal terms which have doubtful meanings in common usage, I reverted to envisioning my audience as ordinary users of camera accessories, and therefore would not have any interest in legal terms. After reading many sample specifications written perhaps by lawyers, the temptation to use legal terms tended to be very enticing. At first I felt that without being legal sounding, the USPTO will not grant a patent to an invention. The notion was incorrect as it turned out in my case.

Additionally, I saw to it that the specification has enough details of sufficient clarity that would enable a person having ordinary skill in the art (PHOSITA), to build the invention. Having some sort of mission like this simplified writing the specification.

The requirements for a specification are clearly identified in the heading "Non-Provisional Application – Specifications – Provisions that apply". The specification for the Digital Camera Lens Guard and Use Extender closely matched the requirements. There are embellishments to the specification however that I felt I should mention.

When in doubt about whether to use standard or English method of measurement, I used standard with English conversion. I was being careful to use approximate instead of exact measurements to avoid limiting the scope of the proposed invention.

I believed that writing the specification should be done at the same time that prototyping the proposed invention was in progress, the reason being that while prototyping the proposed invention, the steps were clearer, more defined, and easily documented. The drawings as well were clearly workable during prototyping.

For clarity, proper markings of the specification were imposed; punctuations were as accurate as possible. Paragraphs and subparagraphs were clearly identified by proper indentions. Capitalizations and bolding where I thought might help clarify the points in the specification, were distinctly shown.

The specification required the inclusion of a claim or claims. In my initial filing, I included one independent claim and nine dependent claims. These are described in the "Writing Claims" heading of this book.

**Specification**

**Title, prior art**

Title –
highlighted

Prior Art

Patent Application of Roy Lique for
"Digital Camera Lens Guard and Use Extender"
Page 1

Cross Reference to Related Applications

This application claims the benefit of provisional patent application Ser. No. 61/849,349, filed 01/25/2013 by the present inventor.

Background Prior Art

The following is a tabulation of some prior arts that presently appear barely relevant to the embodiments of the digital camera lens guard and use extender.

**U.S. Patents**

| Patent Number | Title | Issue Date | Patentee |
|---|---|---|---|
| 8,111,984 | Matte box assembly | February 7, 2012 | Wood; Dennis |
| 8,054,545 | Lens hood for a camera lens | November 8, 2011 | Cheng; Ming-Chung |
| 7,813,639 | Camera cover | October 12, 2010 | Yoneji; Osamu |
| 6,243,540 | Lens barrel assembly | June 5, 2001 | Kume; Hideaki |
| 5,126,881 | Lens hood for a photographic lens | June 30, 1992 | Crema; Rolf |
| 4,533,212 | Accessory holding device for optical instrument | August 6, 1985 | Shimizu; Seiichi |
| 4,384,767 | Clamping device for camera accessory or hood | May 24, 1983 | Kawai; Tohru |
| 4,295,706 | Combined lens cap and sunshade for a camera | October 20, 1981 | Frost; George H. |
| 4,137,540 | Camera matte box | January 30, 1979 | Curtis; Jack |

# Prior art, prior art not adequate

**Prior Art, continued**

**Why prior art is not adequate for the purpose intended by the proposed invention**

Patent Application of Roy Lique for
"Digital Camera Lens Guard and Use Extender"
Page 2

| 3,909,107 | Hood for the lens of optical instruments with pivotally mounted lens cover | September 30, 1975 | Numbers; Jody L. |
| 3,840,883 | CAMERA LENS HOOD | October 8, 1974 | Choate; J. Robert |
| 3,614,196 | COMBINED LENS HOOD AND FILTER SUPPORT | October 19, 1971 | Schlapp; Werner |

In the absence of significant relevance between prior arts and the embodiments of the digital camera lens guard and use extender, the use and benefits of the latter are discussed to create and present a new line of products principally dedicated to digital cameras.

The embodiments of the digital camera lens guard and use extender are in the field of cameras. More particularly, the embodiments extend and expand the capacities of a digital camera while also protecting the camera's lens. Capacities expansion is done with older as well as newer hardware, or a combination of both. Where applicable, expansion will also be done with software.

While the camera market is flooded with new camera models every year, it is also being depleted by obsolescence due to incompatibility with new software and hardware. It is also depleted by mere dislikes of older models and the accessories that come with them. Phone cameras also diminish the popularity of digital cameras. Some victims of obsolescence include multi-image lenses, fisheye lenses, macro and telephoto lenses, square-shaped filters, over-sized and under-sized filters, fog and snow lenses, so on so forth.

Interchangeability of accessories between the digital camera and certain models of expensive cameras, is rare. The digital camera's lack of threads accounts for threaded accessories being almost exclusive monopolies of certain camera models. Some exciting photographs are taken using threaded accessories. Making the interchangeability problem more obvious is the fact that threads come in either metric or English.

**Prior art not adequate, summary**

Why prior art is not adequate for the purpose intended by the proposed invention, continued

Summary

Patent Application of Roy Lique for
"Digital Camera Lens Guard and Use Extender"
Page 3

Notwithstanding the additional sophisticated features coming with new cameras, there are still missed capacities that one would like to have in the digital camera. Examples are documenting an event happening too far from the viewer, or a past time too dangerous for the viewer to come close to the subject.

The additional new features that come with the later digital camera models include expensive electronics that need more protection. The entry of dust, moisture, and other contaminations into the lens area must be minimized in order to maintain the digital camera in an efficient working order. By engaging an embodiment of the digital camera lens guard and use extender to the digital camera and screwing in a camera filter, ample protection of the camera electronics is already provided.

The digital camera equipped with the embodiment of the digital camera lens guard and use extender, meets both the camera users' needs for additional camera capacities and lens and electronics protection without investing in expensive equipments.

Summary

Functionally, the digital camera equipped with an embodiment of the digital camera lens guard and use extender compares with the more expensive models. The embodiment concentrates on enhancing the capacities of the digital camera and protecting its electronics.

With different size adapter rings at the ends of the primary ring of the embodiment, a significant number of camera filters can be used. Obsolete and old camera filters are given new life because they can be used again. With proper adaptations, lenses from other camera models may now be used with the digital camera.

In conjunction with telescopes, binoculars, microscopes and other sighting devices, the digital camera can serve as a tool in industrial applications. More photo opportunities become available for the digital camera user because of the expanded capacities provided by sighting devices.

Aside from being used as industrial tool, the digital camera can now be used in other

**Summary, drawing different views**

Summary,
continued

Drawings,
different views

Patent Application of Roy Lique for
"Digital Camera Lens Guard and Use Extender"
Page 4

applications like law enforcement, crowd monitoring, event reporting, safety applications, research, and fast sports. Moreover, the embodiment of the digital camera lens guard and use extender can be used on different camera brands and models. Consequently, investment in expensive equipments is postponed or skipped entirely for a while.

The digital camera is equipped with sensitive electronics. The chance of the digital camera being damaged due to the entry of dust, moisture, and other contaminations into the lens area is minimized with the use of the embodiment of the digital camera lens guard and use extender. Physical damage to the digital camera due to being dropped, bumped, and hit is also minimized. The digital camera remains portable despite the addition of the ring that comes with the camera mount assembly.

Drawings – Figures

Fig. 1 shows a lens guard with slip type camera mount assembly and a coupler ring, positioned between a camera and a sighting device.

Fig. 2 shows a lens guard with slip type camera mount assembly, positioned between a camera and a camera filter or a sighting device.

Fig. 3 shows a lens guard with screw type camera mount assembly and a coupler ring, positioned between a camera and a sighting device.

Fig. 4 shows a lens guard with screw type camera mount assembly, positioned between a camera and a camera filter or a sighting device.

Fig. 5 shows a slip type barrel track directly securing a camera filter or a sighting device, positioned between a camera and a sighting device.

Fig. 6 shows a screw type barrel track directly securing a camera filter or a sighting device, positioned between a camera and a sighting device.

Fig. 7 is an end view of a primary ring with a stack of outer and inner adapter rings.

## Drawing reference numbers, glossary

**Drawings, reference numerals**

**Glossary of terms**

Patent Application of Roy Lique for
"Digital Camera Lens Guard and Use Extender"
Page 5

Drawings – Reference Numerals

| | |
|---|---|
| 10 – anchor ring | 12 – anchor hole |
| 14 – anchor nut | 16 – anchor screw |
| 18 – base adapter ring | 20 – track mounting ring |
| 22 – barrel track | 24 – primary ring |
| 26 – recess | 28 – outer adapter ring |
| 30 – inner adapter ring | 32 – retaining holes |
| 34 – retaining screws | 36 – insert holes |
| 38 – coupler ring | 40 – track end ring |

Glossary

Before the embodiments of the digital camera lens guard and use extender are described, some terms used here need to be defined.

The term "lens guard" refers to an embodiment of the digital camera lens guard and use extender.

The terms "slip type" and "screw type" refer to the action of connecting one component of the digital camera lens guard and use extender to another.

The terms "digital camera(s)" as used in these specifications, refers to a camera designed to be aimed to shoot pictures of optimal color, with ease and automatic adjustments of shutter speed, aperture, focus, and light sensitivity.

The terms "approximate" and "approximately" apply to numeric values and mean close to being exact.

The term "eyepiece" refers to the lens or lens group closest to the eye in an optical instrument.

# Glossary, main embodiment description, spec location 1

**Glossary of terms, continued**

**Detailed description of the main embodiment of the invention**

---

Patent Application of Roy Lique for
"Digital Camera Lens Guard and Use Extender"
Page 6

The term "sighting device(s)" refers to a device used to assist in aligning or aiming weapons, surveying instruments, or other items by sight.

The term "wall" refers to the surface between the outside and inside diameters of a tube or cylinder.

The term "adapter ring(s)" refers to connectors for joining parts or devices having different sizes and designs, enabling them to be mated, fitted, and work together.

The term "camera filter(s)" refers to circular lens screen of plain or dyed gelatin or glass for controlling the rendering of color or for lessening the intensity of light and for protecting the camera lens.

DETAILED DESCRIPTIONS – FIG. 1 and FIG. 2 – FIRST EMBODIMENTS

Construction of the embodiments of the digital camera lens guard and use extender described in these specifications is based on the camera barrel size range of 40mm to 45mm. All measurements, values, and dimensions are relative to this particular barrel size range. Other size ranges can be used provided their relationships of measurements, values, and dimensions are progressively and proportionally maintained.

Referring now to Fig. 1 and Fig. 2, there are shown embodiments of a digital camera lens guard and use extender having a camera mount assembly, a main housing assembly, and additionally in the case of Fig. 1, a coupler ring 38. The figures further show the position of each embodiment in relation to the digital camera represented by boxed "Digital Camera", and to the attachments represented by boxed "Sighting Device" or boxed "Camera Filter/Sighting Device". When used with the coupler ring 38, the embodiment offers only a single attachment and its position is shown in Fig. 1. With multiple allowable attachments attached one at a time, the position of the embodiment is shown in Fig. 2. The only difference between the embodiments shown in Fig. 1 and Fig. 2 is the presence or absence of the coupler ring 38.

Descriptions of the functions and construction details of the shown embodiments of the digital camera lens guard and use extender follow.

---

## Main embodiment description, spec location 2

Detailed
description of
the main
embodiment of
the invention,
continued

Patent Application of Roy Lique for
"Digital Camera Lens Guard and Use Extender"
Page 7

**Camera mount assembly.** Still referring to Fig. 1 and Fig. 2, an anchor ring 10 of the camera mount assembly is shown as the base that secures the embodiment of the digital camera lens guard and use extender to the digital camera. It is also the basis for most of the sizes, values, measurements, and dimensions of the other components of the embodiment.

**Slip type camera mount assembly.** In more details, still referring to Fig. 1 and Fig. 2, a slip type camera mount assembly comprises the anchor ring 10, an anchor hole 12, an anchor nut 14, an anchor screw 16, and a barrel track 22.

The action of slipping the barrel track 22 into the anchor ring 10 after the anchor ring 10 is mounted on the digital camera, is what "slip type" camera mount assembly refers to. The action, simultaneous with slipping the other end of the barrel 22 to the primary ring 24, secures the embodiment of the digital camera lens guard and use extender to the digital camera.

**Anchor ring 10.** The cylindrical anchor ring 10 of approximately 13mm (0.512") in length, with approximately 44.7mm (1.759") inside diameter, and with approximately 1.5mm (0.058") wall is cut from 6061 grade aluminum. The wall provides sufficient thickness for attaching the anchor ring 10 upright to the digital camera concentrically with the camera barrel. The anchor hole 12 facilitates the attachment of a pair of sufficiently large anchor screw 16 and anchor nut 14, to the anchor ring 10. The anchor screw 16 and the anchor nut 14 are used to secure the barrel track 22.

**Barrel track 22.** The cylindrical barrel track 22 is also from 6061 grade aluminum, cut to an approximate length of 25.4mm (1.0"). Approximately one half of its entire length is inserted into the anchor ring 10 to support a secure connection. The other approximate half serves as a flange that inserts into the primary ring 24. The depth of insertion of the flange into the primary ring 24 is adjustable, making it useful in minimizing the formation of circles around pictures taken by the digital camera.

**Main housing assembly.** In more details still referring to Fig. 1 and Fig. 2, the main housing assembly comprises a cylindrical primary ring 24 with appurtenances. It is in the main housing

137

## Main embodiment description, spec location 3

Detailed
description of
the main
embodiment of
the invention,
continued

Patent Application of Roy Lique for
"Digital Camera Lens Guard and Use Extender"
Page 8

assembly that the major components of the embodiment of the digital camera lens guard and use extender come together and form a secure connection. Photographing activities take place in the main housing assembly.

**Primary ring 24.** The primary ring 24 accepts and secures a camera filter or a sighting device through the embedded inner adapter ring 30 (not shown) or outer adapter ring 28 at its ends. Additionally in Fig. 1, the primary ring 24 accepts and secures a sighting device through a coupler ring 38.

The primary ring 24 is also cut from 6061 grade aluminum to approximately 25.4mm (1.0") long. Sufficient space for embedding the commonly used size ranges of adapter rings is provided by its approximate outside diameter of 63.5mm (2.5") and approximate wall of 6.4mm (0.252"). Its approximate inside diameter of 44.5mm (1.752") provides the additional track for the camera barrel to extend and retract without obstruction.

Owing to the identical ends of the primary ring 24 with different size adapter rings, the directions at which the ends point are reversible, offering more opportunities for attaching different camera filters or sighting devices, one at a time.

**Adapter rings.** Size-matched and gender-matched camera filter or sighting device is accepted at either end of the primary ring 24. The embedded adapter rings at the ends of the primary ring 24 immediately provide more opportunities to use different camera filters and sighting devices, one at a time. Since circles around images taken by the digital camera are sometimes caused by the addition of loose adapter rings and camera filters, care is observed that only enough of them are used as needed.

The inside and outside diameters of a target adapter ring are modified as necessary to sufficient sizes so that the adapter ring can be embedded at either end of the primary ring 24. Optionally, its male threads are stripped off. Two types of adapter rings are optionally embedded at either end of the primary ring 24, namely, inner adapter ring 30 and outer adapter ring 28. Their construction details are described as follows:

**Inner adapter ring 30.** To attach the inner adapter ring 30 at either end of the

## Main embodiment description, spec location 4

Detailed
description of
the main
embodiment of
the invention,
continued

Patent Application of Roy Lique for
"Digital Camera Lens Guard and Use Extender"
Page 9

primary ring 24, an approximately 6.4mm (0.252") deep recess 26 with sufficient circumference to accept the target inner adapter ring 30, is carved. The inner adapter ring 30 is attached resting at the bottom of the recess 26 with the female threads oriented outwards, using industry grade adhesive.

**Outer adapter ring 28.** To attach the outer adapter ring 28 at either end of the primary ring 24, an optional approximately 3.2mm (0.125") deep recess 26 with sufficient circumference to accept the target outer adapter ring 28, is carved. The outer adapter ring 28 is attached resting at the bottom of the recess 26 with the female threads oriented outwards, using industry grade adhesive.

An alternative way to embed the outer adapter ring 28 into the primary ring 24 is to cleanly cut off the recessed portion of the primary ring 24, referring to previous paragraph. Using industry grade adhesive, the outer adapter ring 28 is inserted and attached flushed with the end of the cut off portion. The cut off portion is attached back to the primary ring 24 making sure the female threads are oriented outwards.

The outer adapter ring 28 can also be attached directly to either end of the primary ring 24 without carving the recess 26. It only needs to be stripped off of its male threads and attached to the primary ring 24 with the female threads oriented outwards, using industry grade adhesive.

**Stack of adapter rings.** Building a stack of adapter rings as shown in Fig. 7, follows the steps described for embedding the inner adapter ring 30 and the outer adapter ring 28. The inner adapter rings 30 progressively get smaller as the stack grows, each inner adapter ring 30 resting approximately 6.4mm (0.252") deeper from the preceding larger one.

**Retaining holes 32 and retaining screws 34.** In order to provide ample handling surface, retaining holes 32 of sufficient size are drilled approximately 6.4mm (0.252") from either end of the primary ring 24. Because at least one retaining screw 34 is used to secure the barrel track 22 and another one to secure the coupler ring 38, the locations of

**Main embodiment description, spec location 5**

Detailed
description of
the main
embodiment of
the invention,
continued

Patent Application of Roy Lique for
"Digital Camera Lens Guard and Use Extender"
Page 10

the retaining holes 32 are sufficiently far apart to allow the retaining screws 34 to turn freely.

**Insert holes 36.** The optional insert holes 36 are of sufficient size to accept inserts for expansion and improvements. They are drilled onto the primary ring 24 at desired locations that do not interfere with the outer adapter rings 28, inner adapter rings 30, and retaining screws 34.

**Coupler ring 38.** In more details referring to Fig. 1, the coupler ring 38 provides the connection between the main housing assembly and the sighting device. The sighting device is secured to the main housing assembly by properly inserting its eyepiece into the coupler ring 38 and the coupler ring 38 into the primary ring 24.

The coupler ring 38 is of sufficient size and length, preferably between 25.4mm (1.0") and 50.8mm (2.0"), and has an outside diameter closely matching the inside diameter of the primary ring 24. The inside diameter of the coupler ring 38 varies depending on the size of the target eyepiece of the sighting device. If necessary, the coupler ring 38 is machined to alter its inside diameter in order to match it with the size of the eyepiece of the sighting device. The coupler ring 38 is also cut from 6061 grade aluminum.

The construction details of the embodiments of the digital camera lens guard and use extender as shown in Fig. 1 and Fig. 2 are that the embodiments may be made of metal or of any other sufficiently rigid and strong material such as high-strength plastic and the like. Further, the various components of the embodiments of the digital camera lens guard and use extender can be made of different materials from different sources, brands, and styles.

Operation – Fig. 1 and Fig. 2

Referring to Fig. 1 and Fig. 2, either end of the barrel track 22 is inserted into the anchor ring 10. Anchor screw 16 is tightened to secure the barrel track 22.

The open end of the barrel track 22 is inserted into either end of the primary ring 24. It is inserted opposite the end where the camera filter or sighting device is attached. The size

## Main embodiment description, spec location 6

Detailed description of the main embodiment of the invention, continued

Patent Application of Roy Lique for
"Digital Camera Lens Guard and Use Extender"
Page 11

and type of camera filter or sighting device determine which end of the primary ring 24 needs to accept and secure the barrel track 22.

Insertion depth of the barrel track 22 into the primary ring 24 is fixed for the distance the camera barrel has to extend. One or more retaining screws 34 are tightened to secure the barrel track 22.

In more details referring to Fig. 1, the eyepiece of a sighting device is inserted into either end of the coupler ring 38. If the outside circumference of the eyepiece is smaller than the inside circumference of the coupler ring 38, a fitting (not shown) is used to make the insertion snugly secure. The open end of the coupler ring 38 is inserted as far as it can go or until it is stopped by the barrel track 22, into the primary ring 24.

Alternatively, the use of the coupler ring 38 can be done away with as shown in Fig. 2. Size-matched and gender-matched camera filter or sighting device is screwed directly to either the outer adapter ring 28 or the inner adapter ring 30 (not shown). If necessary, loose adapter rings are added to find a match between the embedded adapter rings and the target camera filter or sighting device.

In more details referring to Fig. 1 and Fig. 2, as shown, the embodiments of the digital camera lens guard and use extender provide for quick and easy installation of the camera filter or the sighting device to the main housing assembly. They also facilitate their quick and easy mounting and dismounting to and from the digital camera. More size-matched and gender-matched camera filters and sighting devices unusable with the digital camera before, become available now with the use of the outer adapter rings 28 and the inner adapter rings 30.

Fig. 3 and Fig. 4 – Additional Embodiments

Referring now to Fig. 3 and Fig. 4, there are shown embodiments of the digital camera lens guard and use extender having a camera mount assembly, a main housing assembly, and additionally in the case of Fig. 3, a coupler ring 38. The figures further show the position of each embodiment in relation to the digital camera represented by boxed "Digital Camera",

**Additional embodiment description, page 1**

Detailed
description of
additional
embodiment
of the invention

Patent Application of Roy Lique for
"Digital Camera Lens Guard and Use Extender"
Page 12

and to the attachments represented by boxed "Sighting Device" or boxed "Camera Filter/Sighting Device". When used with the coupler ring 38, the embodiment offers only a single attachment and its position is shown in Fig. 3. With multiple allowable attachments attached one at a time, the position of the embodiment is shown in Fig. 4. The only difference between the embodiments shown in Fig. 3 and Fig. 4 is the presence or absence of the coupler ring 38.

Descriptions of the functions and construction details of the shown embodiments of the digital camera lens guard and use extender follow.

**Camera mount assembly.** Still referring to Fig. 3 and Fig. 4, a base adapter ring 18 of the camera mount assembly secures the embodiment of the digital camera lens guard and use extender to the digital camera. It is the basis for most of the sizes, values, measurements, and dimensions of the other components of the embodiments.

**Screw type camera mount assembly.** In more details still referring to Fig. 3 and Fig. 4, the screw type camera mount assembly comprises the modified base adapter ring 18 and the configured barrel track 22.

The action of screwing the track mounting ring 20 to the modified base adapter ring 18 after the base adapter ring 18 is mounted on the digital camera, is what "screw type" camera mount assembly refers to. The action, simultaneous with slipping the other end of the barrel 22 to the primary ring 24, secures the embodiment of the digital camera lens guard and use extender to the digital camera.

**Base adapter ring 18.** Initially, the inside diameter of the base adapter ring 18 is approximately 40mm (1.575") to 45mm (1.772") and its outside diameter is approximately 50mm (1.969"). The sizes are suitable for modification by machining, to circumferentially enclose the camera barrel. With its male threads stripped off, the base adapter ring 18 is attached to the digital camera upright concentrically with the camera barrel and with the female threads oriented outwards, using industry grade adhesive. To allow the camera barrel to extend and retract without obstruction, a space is maintained between the

## Additional embodiment description, page 2

Detailed description of additional embodiment of the invention, continued

Patent Application of Roy Lique for
"Digital Camera Lens Guard and Use Extender"
Page 13

camera barrel and the base adapter ring 18.

**Barrel track 22.** The barrel track 22 is also cut from 6061 grade aluminum to an approximate length of 25.4mm (1.0"). Using industry grade adhesive, one end is fitted with the modified track mounting ring 20 which is size-matched and gender-matched with the base adapter ring 18. The track mounting ring 20 is made flush with the end of the barrel track 22. The other end serves as a flange that inserts into the primary ring 24. The depth of insertion of the flange into the primary ring 24 is adjustable, making it useful in minimizing the formation of circles around pictures taken by the digital camera.

**Main housing assembly.** The main housing assembly is identical to that described in the embodiments shown in Fig. 1 and Fig. 2.

**Coupler ring 38.** The coupler ring 38 is identical to that described in the embodiments shown in Fig. 1 and Fig. 2.

The construction details of the embodiments of the digital camera lens guard and use extender as shown in Fig. 3 and Fig. 4 are that the embodiments may be made of metal or of any other sufficiently rigid and strong material such as high-strength plastic and the like. Further, the various components of the embodiments of the digital camera lens guard and use extender can be made of different materials from different sources, brands, and styles.

Operation – Fig. 3 and Fig. 4

Referring to Fig. 3 and Fig. 4, the track mounting ring 20 is screwed into the base adapter ring 18.

The open end of the barrel track 22 is inserted into either end of the primary ring 24. It is inserted opposite the end where the camera filter or sighting device is attached. The size and type of camera filter or sighting device determine which end of the primary ring 24 needs to accept and secure the barrel track 22.

Insertion depth of the barrel track 22 into the primary ring 24 is fixed for the distance the camera barrel has to extend. One or more retaining screws 34 are tightened to secure the

**Alternative embodiment description**

Detailed
description of
additional
embodiment of
the invention,
continued

Detailed
description of
alternative
embodiment of
the invention

Patent Application of Roy Lique for
"Digital Camera Lens Guard and Use Extender"
Page 14

barrel track 22.

In more details referring to Fig. 3, the eyepiece of a sighting device is inserted into either end of the coupler ring 38. If the outside circumference of the eyepiece is smaller than the inside circumference of the coupler ring 38, a fitting (not shown) is used to make the insertion snugly secure. The open end of the coupler ring 38 is inserted as far as it can go or until it is stopped by the barrel track 22, into the primary ring 24.

Alternatively, the use of the coupler ring 38 can be done away with as shown in Fig. 4. Size-matched and gender-matched camera filter or sighting device is screwed directly to either the outer adapter ring 28 or the inner adapter ring 30 (not shown). If necessary, loose adapter rings are added to find a match between the embedded adapter rings and the target camera filter or sighting device.

In more details referring to Fig. 3 and Fig. 4, as shown, the embodiments of the digital camera lens guard and use extender provide for quick and easy installation of the camera filter or the sighting device to the main housing assembly. They also facilitate their quick and easy mounting and dismounting to and from the digital camera. More size-matched and gender-matched camera filters and sighting devices unusable with the digital camera before become available now with the use of the outer adapter rings 28 and the inner adapter rings 30.

Fig. 5 and Fig. 6 – Alternative Embodiments

Referring to Fig. 5 and Fig. 6, there are shown simplified embodiments of the digital camera lens guard and use extender. The figures further show the position of each embodiment in relation to the digital camera represented by boxed "Digital Camera". With multiple allowable attachments represented by boxed "Camera Filter/Sighting Device" attached one at a time, the relative positions of the embodiments are also shown in both figures.

In Fig. 5 the end of the barrel track 22 opposite the end that slips into the anchor ring 10 is fitted with a track end ring 40 with the female threads oriented outwards. In Fig. 6, fitting is

**Advantages of proposed invention**

Detailed
description of
alternative
embodiment of
the invention,
continued

Advantages of
using the
proposed
invention

Patent Application of Roy Lique for
"Digital Camera Lens Guard and Use Extender"
Page 15

done at the end opposite that which screws into the base adapter ring 18. In both instances, fitting is done using industry grade adhesive and ensuring that the track end ring 40 is flushed with the end of the barrel track 22. The camera filter or sighting device is screwed directly into the track end ring 40.

The construction details of the embodiments of the digital camera lens guard and use extender as shown in Fig. 5 and Fig. 6 are that the embodiments may be made of metal or of any other sufficiently rigid and strong material such as high-strength plastic and the like. Further, the various components of the embodiments of the digital camera lens guard and use extender can be made of different materials from different sources, brands, and styles.

Referring to Fig. 5 and Fig. 6, as shown, the embodiments of the digital camera lens guard and use extender provide for simple and direct use of camera filters and sighting devices by completely bypassing the main housing assembly.

Advantages

Broadly, from the description above, a number of advantages of most embodiments of the digital camera lens guard and use extender become evident:

(a) The embodiments of the digital camera lens guard and use extender have the advantage of possibly being one of the few dedicated to digital cameras.

(b) Due to its simple design, future changes on the embodiments of the digital camera lens guard and use extender will be easily implemented.

(c) The features added by sighting devices make the digital camera more adaptable to various photo opportunities.

(d) With the added features of a sighting device, the digital camera can be used as an industrial tool.

(e) Mounting and dismounting of an embodiment to and from the digital camera takes only few turns of the track mounting ring or the anchor screw.

## Conclusion, ramifications, and scope

**Advantages of using the proposed invention, continued**

**Conclusion, ramifications, and scope**

Patent Application of Roy Lique for
"Digital Camera Lens Guard and Use Extender"
Page 16

(f) The digital camera remains portable despite the addition of a base ring.

(g) Embodiments of the digital camera lens guard and use extender of different sizes can be manufactured for different digital camera types, models, and sighting devices, provided appropriate matching adapter rings are used.

(h) The digital cameras are now able to take pictures previously possible only with the more expensive cameras.

(i) Camera protection extends the life of the digital camera and is accomplished with just a few turns of a camera filter.

(j) Camera users will benefit from innovations and improvements from two different industrial classifications, namely, digital cameras and sighting devices.

(k) With the different size adapter rings fitted at the ends of the primary ring, immediately a large number of camera filters becomes available.

(l) Tripods are used less frequently because of the nature of digital cameras.

Conclusion, Ramifications, and Scope

Accordingly, a digital camera user will see that a digital camera enabled by the embodiments of the digital camera lens guard and use extender is well adapted to various photo taking sessions. As the embodiments are easy to mount and dismount to and from the digital camera, more photographic events can be recorded.

In terms of functionalities, the market availability of camera filters and sighting devices makes the digital camera comparable with the more expensive types and models. With the option to choose which end of the main housing assembly to attach camera filters and sighting devices, the possibilities become more numerous.

More specifically, the following are few examples of the use of the digital camera enabled by the embodiments of the digital camera lens guard and use extender:

## Sample usages of proposed invention

Example usages of the proposed invention

Patent Application of Roy Lique for
"Digital Camera Lens Guard and Use Extender"
Page 17

- surveillance from a distance when getting close endangers the observer,

- observing a phenomenon such as unusual celestial events,

- research such as observing the habits of certain insects,

- crowd observation such as in parades and demonstrations,

- monitoring such as vehicles behaving erratically,

- past time such as bird-watching and whale-watching,

- safaris such as observing wild animals from a distance,

- emergency reporting such as a traffic accident,

- fire fighting such as reporting a fire at a distant canyon,

- progress report such as of a mountain climber,

- fast sports such as tennis matches and basketball games,

- traffic surveillance such as monitoring speeding cars,

- weather observation such as monitoring snow levels,

- crowd safety such as life guarding on the beach, and

- law enforcement such as unruly crowd.

While the foregoing written descriptions of the embodiments of the digital camera lens guard and use extender enables one of ordinary skill to make and use what is considered presently to be the best mode thereof, those of ordinary skills will understand and appreciate the existence of variations, combinations, and equivalents of the specific embodiment, method, and examples herein. The digital camera lens guard and use extender should therefore not be limited by the above described embodiments, methods, and examples, but by all embodiments and methods within the scope and spirit of the digital camera lens guard and use extender as claimed.

## Non-provisional Application – Claims

### Some rules pertaining to claims

PCT Article 6

The Claims

The claim or claims shall define the matter for which protection is sought. Claims shall be clear and concise. They shall be fully supported by the description.

PCT Rule 6

The Claims

6.1 Number and Numbering of Claims

- (a) The number of the claims shall be reasonable in consideration of the nature of the invention claimed.
- (b) If there are several claims, they shall be numbered consecutively in Arabic numerals.
- (c) The method of numbering in the case of the amendment of claims shall be governed by the Administrative Instructions.

6.2 References to Other Parts of the International Application

- (a) Claims shall not, except where absolutely necessary, rely, in respect of the technical features of the invention, on references to the description or drawings. In particular, they shall not rely on such references as: "as described in part ... of the description," or "as illustrated in figure ... of the drawings."
- (b) Where the international application contains drawings, the technical features mentioned in the claims shall preferably be followed by the reference signs relating to such features. When used, the reference signs shall preferably be placed between parentheses. If inclusion of reference signs does not particularly facilitate quicker understanding of a claim, it should not be made. Reference signs may be removed by a designated Office for the purposes of publication by such Office.

6.3 Manner of Claiming

- (a) The definition of the matter for which protection is sought shall be in terms of the technical features of the invention.
- (b) Whenever appropriate, claims shall contain:

- (i) a statement indicating those technical features of the invention which are necessary for the definition of the claimed subject matter but which, in combination, are part of the prior art,
- (ii) a characterizing portion - preceded by the words "characterized in that," "characterized by," "wherein the improvement comprises," or any other words to the same effect - stating concisely the technical features which, in combination with the features stated under (i), it is desired to protect.

(c) Where the national law of the designated State does not require the manner of claiming provided for in paragraph (b), failure to use that manner of claiming shall have no effect in that State provided the manner of claiming actually used satisfies the national law of that State.

## 6.4 Dependent Claims

- (a) Any claim which includes all the features of one or more other claims (claim in dependent form, hereinafter referred to as "dependent claim") shall do so by a reference, if possible at the beginning, to the other claim or claims and shall then state the additional features claimed. Any dependent claim which refers to more than one other claim ("multiple dependent claims") shall refer to such claims in the alternative only. Multiple dependent claims shall not serve as a basis for any other multiple dependent claims. Where the national law of the national Office acting as International Searching Authority does not allow multiple dependent claims to be drafted in a manner different from that provided for in the preceding two sentences, failure to use that manner of claiming may result in an indication under Article 17(2)(b) in the international search report. Failure to use the said manner of claiming shall have no effect in a designated State if the manner of claiming actually used satisfies the national law of that State.
- (b) Any dependent claim shall be construed as including all the limitations contained in the claim to which it refers or, if the dependent claim is a multiple dependent claim, all the limitations contained in the particular claim in relation to which it is considered.
- (c) All dependent claims referring back to a single previous claim, and all dependent claims referring back to several previous claims, shall be grouped together to the extent and in the most practical way possible.

## 6.5 Utility Models

Any designated State in which the grant of a utility model is sought on the basis of an international application may, instead of Rules 6.1 to 6.4, apply in respect of the matters regulated in those Rules the provisions of its national law concerning utility models once the processing of the international application has started in that State, provided that the applicant shall be allowed at least two months from the expiration of the time limit applicable under Article 22 to adapt his application to the requirements of the said provisions of the national law.

PCT Administrative Instructions Section 205.

Numbering and Identification of Claims Upon Amendment

- (a) Amendments to the claims under Article 19 or Article 34(2)(b) may be made either by cancelling one or more entire claims, by adding one or more new claims or by amending the text of one or more of the claims as filed. Where a claim is cancelled, no renumbering of the other claims shall be required. In all cases where claims are renumbered, they shall be renumbered consecutively in Arabic numerals.
- (b) The applicant shall, in the letter referred to in Rule 46.5(b) or Rule 66.8(c), indicate the differences between the claims as filed and the claims as amended or, as the case may be, differences between the claims as previously amended and currently amended. He shall, in particular, indicate in the said letter, in connection with each claim appearing in the international application (it being understood that identical indications concerning several claims may be grouped), whether:
  - (i) the claim is unchanged;
  - (ii) the claim is cancelled;
  - (iii) the claim is new;
  - (iv) the claim replaces one or more claims as filed;
  - (v) the claim is the result of the division of a claim as filed;
  - (vi) the claim replaces one or more claims as previously amended;
  - (vii) the claim is the result of the division of a claim as previously amended.

37 CFR 1.436 The claims.

The requirements as to the content and format of claims are set forth in PCT Art. 6 and PCT Rules 6, 9, 10 and 11 and shall be adhered to. The number of the claims shall be reasonable, considering the nature of the invention claimed.

The claim or claims must "define the matter for which protection is sought." Claims must be clear and concise. They must be fully supported by the description. PCT Rule 6 contains detailed requirements as to the number and numbering of claims, the extent to which any claim may refer to other parts of the international application, the manner of claiming, and dependent claims. As to the manner of claiming, the claims must, whenever appropriate, be in two distinct parts; namely, the statement of the prior art and the statement of the features for which protection is sought ("the characterizing portion").

The physical requirements for the claims are the same as those for the description. Note that the claims must commence on a new sheet.

The procedure for rectification of obvious mistakes is explained in MPEP § 1836. The omission of an entire sheet of the claims cannot be rectified without affecting the international filing date, except in applications filed on or after April 1, 2007, where, if the application, on its initial receipt date, contained a priority claim and a proper incorporation by reference statement, the original international filing date may be retained if the submitted correction was completely contained in the earlier application. See PCT Rules 4.18 and 20.6. It is recommended that a request for rectification of obvious mistakes in the claims be made only if the mistake is liable to affect the international search; otherwise, the rectification should be made by amending the claims.

The claims can be amended during the international phase under PCT Article 19 on receipt of the international search report, during international preliminary examination if the applicant has filed a Demand, and during the national phase.

Multiple dependent claims are permitted in international applications before the United States Patent and Trademark Office as an International Searching and International Preliminary Examining Authority or as a Designated or Elected Office, if they are in the alternative only and do not serve as a basis for any other multiple dependent claim (PCT Rule 6.4(a), 35 U.S.C. 112). The claims, being an element of the application, should start on a new page (PCT Rule 11.4). Page numbers must not be placed in the margins (PCT Rule 11.7(b)). Line numbers should appear in the right half of the left margin (PCT Rule 11.8(b)). See PCT Rule 11.6(e).

The number of claims shall be reasonable, considering the nature of the invention claimed (37 CFR 1.436).

## How I developed initial claims

Writing claims was one of the harshest parts of securing a patent for an invention as I immediately realized after filing the Non-provisional Application. Even with repeated references to the "Patent It Yourself" book by Attorney David Pressman, the task was daunting to me owing to my lack of experience.

**Original**

I claim:

1. a digital camera lens guard and use extender for adding deviant picture taking capacities to a digital camera, and protecting said camera's lens and electronics, comprising,

   a. a camera mount assembly having a circular base and a cylindrical ring of sufficient size and length, for providing enclosure and track for said camera's barrel,

   b. a main housing assembly having a cylindrical ring of sufficient size and length with a plurality of retaining screws and adapter rings, for accepting and securing matching camera filters and sighting devices.

2. the digital camera lens guard and use extender of claim 1, wherein said camera mount assembly having a locking method, for engaging said circular base and said cylindrical ring together.

3. the digital camera lens guard and use extender of claim 2 wherein said locking method is of slip type.

4. the digital camera lens guard and use extender of claim 2 wherein said locking method is of screw type.

5. the digital camera lens guard and use extender of claim 1 wherein said cylindrical ring of said camera mount assembly being fitted with an adapter ring, for securing camera filters or sighting device.

6. the digital camera lens guard and use extender of claim 1 wherein said cylindrical ring of said main housing assembly having more than one embed locations, for securing said adapter rings of said main housing assembly.

7. the digital camera lens guard and use extender of claim 6 wherein said embed location is below the surface at the end of said cylindrical ring.

8. the digital camera lens guard and use extender of claim 6 wherein said embed location is on the surface at the end of said cylindrical ring.

9. the digital camera lens guard and use extender of claim 6 wherein said embed location is in-between a cut off portion of said cylindrical ring and said cylindrical ring itself,

10. the digital camera lens guard and use extender of claim 1 wherein said cylindrical ring of said main housing assembly being inserted with a coupler ring, for securing the eyepiece of a sighting device,

whereby the digital camera lens guard and use extender adds picture taking capacities that are not normally common with a digital camera; provides protection for the camera's lens and electronics; facilitates the use of camera filters, adapter rings, and sighting devices; enables the digital camera to take pictures normally; enables the camera to take pictures of images formed at the eyepiece of a sighting device; and prepares itself for future changes and improvements.

**Final**

While the foregoing written descriptions of the embodiments of the digital camera lens guard and use extender enables one of ordinary skill to make and use what is considered presently to be the best mode thereof, those of ordinary skills will understand and appreciate the existence of variations, combinations, and equivalents of the specific embodiment, method, and examples herein. The digital camera lens guard and use extender should therefore not be limited by the above described embodiments, methods, and examples, but by all embodiments and methods within the scope and spirit of the digital camera lensoguard and use extender as claimed.

I claim:

1. A digital camera lens guard and use extender for adding deviant picture-taking capacities to a digital camera, and protecting said camera's lens, barrel, and electronics, comprising:

   a. a camera mount assembly having a circular base and a cylindrical ring of sufficient size and length and a plurality of locking means, for allowing unobstructed movement of said digital camera's barrel,

   b. a main housing assembly having a primary ring of sufficient size and length with a plurality of retaining screws and embed locations for adapter rings on each end, for union with camera filters and sighting devices.

2. The digital camera lens guard and use extender of claim 1 wherein said embed location is a circular recess carved approximately one-eighth to one-fourth inch at each said end of said primary ring of said main housing assembly.

3. The digital camera lens guard and use extender of claim 1 wherein said primary ring of said main housing assembly being inserted with a coupler ring, for securing an eyepiece of a sighting device,

whereby the digital camera lens guard and use extender adds picture-taking capacities to said digital camera beside that of point-and-shoot method; provides protection for said digital camera's lens, barrel, and electronics; facilitates usage of camera filters, adapter rings, and sighting devices; enables said digital camera to take pictures of images formed at said eyepiece of a sighting device; and allows said digital camera to accept new picture-taking capacities that are introduced in the form of adapter rings.

I had to be very brief and clear in describing my invention in **[readers, please note]** a single sentence that allows only one capital, one period, with no dashes, quotations, parentheses, and abbreviations. With the sentence I should be able to describe the invention in as broad as possible manner to cover all its aspects that can be infringed upon. The infringement I am referring to is not confined to the particular industry of my invention and region where it is located, but also to other industries and regions including countries that have patent rights arrangement with the United States.

The writing must conform to the rigid format prescribed by the USPTO, which is undoubtedly very exacting. Independent and dependent claims must be depicted by proper indentions and use of punctuations to indicate and visually establish the relationship. Though I slightly came up lacking in fully understanding the use of antecedents, with their use, I was able to construct trial sentences that looked decent.

After few attempts, I was very proud to come up with a set comprising one independent claim and nine dependent claims marked "Original" in the adjoining image. That was the best I could do, my knowledge about writing claims having peaked at that point.

The verdict on the validity and allowability of the claims did not come until the prosecution of the application started several months later with the release of the first office action. As it turned out, the USPTO examiner found several defects in the claims. Ultimately, the number of claims was reduced from ten to three. The three claims are marked "Final" in the adjoining image.

Reducing the number of claims from ten to three was not as simple as it appeared to be; it had to undergo through prosecution by the USPTO. How the reduction came about is discussed in the Office Actions heading of the book. Meanwhile, I continued working on the prototype of the invention, the Digital Camera Lens Guard and Use Extender.

# Specification

Claims – Page 1

**Claims, Page 1**

Patent Application of Roy Lique for
"Digital Camera Lens Guard and Use Extender"
Page 18

CLAIMS: I claim:

1. a digital camera lens guard and use extender for adding deviant picture taking capacities to a digital camera, and protecting said camera's lens and electronics, comprising,

    a. a camera mount assembly having a circular base and a cylindrical ring of sufficient size and length, for providing enclosure and track for said camera's barrel,

    b. a main housing assembly having a cylindrical ring of sufficient size and length with a plurality of retaining screws and adapter rings, for accepting and securing matching camera filters and sighting devices,

2. the digital camera lens guard and use extender of claim 1, wherein said camera mount assembly having a locking method, for engaging said circular base and said cylindrical ring together,

    3. the digital camera lens guard and use extender of claim 2 wherein said locking method is of slip type,

    4. the digital camera lens guard and use extender of claim 2 wherein said locking method is of screw type,

5. the digital camera lens guard and use extender of claim 1 wherein said cylindrical ring of said camera mount assembly being fitted with an adapter ring, for securing camera filters or sighting devices,

6. the digital camera lens guard and use extender of claim 1 wherein said cylindrical ring of said main housing assembly having more than one embed locations, for securing said adapter rings of said main housing assembly,

    7. the digital camera lens guard and use extender of claim 6 wherein said embed location is below the surface at the end of said cylindrical ring,

    8. the digital camera lens guard and use extender of claim 6 wherein said

154

Claims, Page 2

Patent Application of Roy Lique for
"Digital Camera Lens Guard and Use Extender"
Page 19

embed location is on the surface at the end of said cylindrical ring,

9.  the digital camera lens guard and use extender of claim 6 wherein said
    embed location is in-between a cut off portion of said cylindrical ring and said
    cylindrical ring itself,

10. the digital camera lens guard and use extender of claim 1 wherein said cylindrical
    ring of said main housing assembly being inserted with a coupler ring, for securing
    the eyepiece of a sighting device,

whereby the digital camera lens guard and use extender adds picture taking capacities that
are not normally common with a digital camera; provides protection for the camera's lens
and electronics; facilitates the use of camera filters, adapter rings, and sighting devices;
enables the digital camera to take pictures normally; enables the camera to take pictures of
images formed at the eyepiece of a sighting device; and prepares itself for future changes
and improvements.

Specification - Object of the Disclosure

**Abstract of the disclosure**

Patent Application of Roy Lique for
"Digital Camera Lens Guard and Use Extender"
Page 20

ABSTRACT OF THE DISCLOSURE

A digital camera lens guard and use extender to: extend the picture-taking capacities of the digital camera by using different camera filters; enable the digital camera to take pictures of the images formed at the eyepiece of a sighting device; protect the camera lens from dust, moisture, and contaminations; make the digital camera usable as an industrial tool; allow the digital camera to be used in photographing activities such as surveillance, sports, past times, research, astronomy; and maintain the original functions of the digital camera.

## Information Disclosure Statement

Patent Search

Specification for a patent requires that a listing of prior arts be included particularly on the first page. In addition, an Information Disclosure Statement, though may not be filed at the same time as the Application Data Sheet, requires the listing as well. Patent search in my case was unavoidable; I had to it.

Patent search requires the familiarization with the following websites:

For Index to the U.S. Patent Classifications:
http://www.uspto.gov/web/patents/classification/uspcindex/indextouspc.htm and
http://www.uspto.gov/go/classification/uspcindex/indextouspc.htm

For Manual of Classification:
http://www.uspto.gov/web/offices/ac/ido/oeip/taf/moc/jobjects/search.htm

For searching the data base of patents: http://patft.uspto.gov

The first thing I did was to secure the proposed invention's class and subclass highlighted in the picture. Owing to the number of pairs that cover different products, I picked the pair that closely matched the description of my proposed invention, effectively reducing the number of searches I had to make.

| ☆ Customize Links | ᕼ Suggested Sites | ☆ Web Slice Gallery |
|---|---|---|

Camera

| | | |
|---|---|---|
| Animated cartoon | 352 / | 87 |
| Carrier or holder | 224 / | 908" |
| Hand | 294 / | 139 |
| Cases | 206 / | 316.2 |
| Cinematographic | D16 / | 205+ |
| Compound lens system with | 359 / | 363 |
| Copy making, eg mechanical negative | 430 / | 951" |
| Copying | 355 / | 18+ |
| Design | D16 / | 200+ |
| Electronic still | 358 / | 909.1" |
| Scanned semiconductor matrix | 348 / | 294+ |
| Exposure control, automatic | 396 / | 213+ |
| Film winding spools | 242 / | 600+ |
| Flash unit | 396 / | 155+ |
| Focusing, automatic | 396 / | 89+ |
| Lens mount | 396 / | 529+ |
| Angularly adjustable camera | 396 / | 342 |
| Front or lens mount | 396 / | 342 |
| Lucida | 359 / | 447 |
| Mammography study | 128 / | 915" |
| Microscope | 359 / | 372+ |
| Motion picture | 352 | |
| Film winding | 242 / | 324.1+ |
| Multiple exposure on single plate | 396 / | 335+ |
| Obscura | 359 / | 448 |
| Panoramic | 396 / | 20+ |
| scope | | 403+ |

| | | |
|---|---|---|
| Submarine motion picture | 352 / | 132 |
| Supports | 396 / | 419+ |
| Telescope | 359 / | 419+ |
| Television | D16 / | 202 |
| Time lapse | 352 / | 84 |
| Toy (design) | D21 / | 514 |
| Tripod | 396 / | 419+ |
| Twin lens reflex | 396 / | 353 |
| Ultrasonic scanning | 600 / | 437 |

Using the class and subclass pair that I just noted, I opened the Patent Full-Text Databases window at: http://patft.uspto.gov. Searching by Class enabled me to display all the patents covered by the class/subclass pair.

Patent Full-Text Databases

The Full-Text Databases window opens the facility for different types of searches, from simple to complex. Although Searching by Class gave me the list of patents that closely matched my proposed invention, I also used Quick Search, Advanced Search, and Number Search, to be absolutely sure that I was not missing obvious patents. The patent examiner might see them

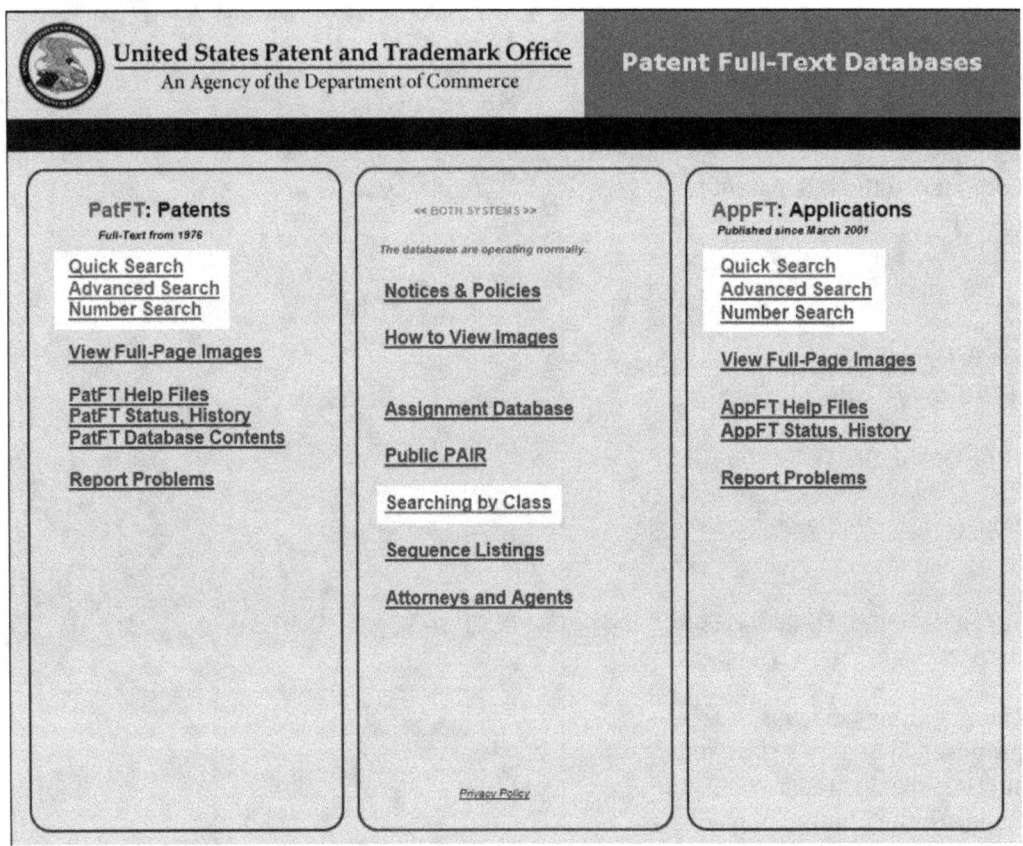

as being purposely hidden from examination to enhance the chance of allowance of my claims. To refine the results of my searches, I stored them in an Excel spreadsheet where I could use additional formulas to further fine tune them if necessary.

Sample search result

## USPTO PATENT FULL-TEXT AND IMAGE DATABASE

| Home | Quick | Advanced | Pat Num | Help |
|---|---|---|---|---|

| Next List | Bottom | View Cart |
|---|---|---|

*Searching US Patent Collection...*

**Results of Search in US Patent Collection db for:**
**camera AND lens**: 138615 patents.
*Hits 1 through 50 out of 138615*

[ Next 50 Hits ]

[ Jump To ] [        ]

[ Refine Search ] camera AND lens

| PAT. NO. | Title |
|---|---|
| 1 RE46,385 **T** | Driving control device and operation device |
| 2 RE46,383 **T** | Deceleration of hadron beams in synchrotrons designed for acceleration |
| 3 9,642,295 **T** | Component mounting device |
| 4 9,642,226 **T** | Digital load control system providing power and communication via existing power wiring |
| 5 9,641,984 **T** | Support of OTDOA positioning using ambiguous cells |
| 6 9,641,920 **T** | Support structures for computing devices |
| 7 9,641,869 **T** | Video camera system with distributed control and methods for use therewith |
| 8 9,641,840 **T** | Processing device and image processing method for encoding and decoding image |
| 9 9,641,838 **T** | Moving image coding apparatus, method and program |
| 10 9,641,826 **T** | System and method for displaying distant 3-D stereo on a dome surface |
| 11 9,641,825 **T** | Gated 3D camera |
| 12 9,641,824 **T** | Method and apparatus for making intelligent use of active space in frame packing format |
| 13 9,641,820 **T** | Advanced multi-band noise reduction |
| 14 9,641,818 **T** | Kinetic object removal from camera preview image |

## Final search result summary

The patents in the Search Results Summary are what appeared on the first page of the specification and ultimately in the final allowance. They are also listed in the Information Disclosure Statement.

As an added convenience resulting from the patent search to fill the requirements for listing

| | A | C | D | E | F |
|---|---|---|---|---|---|
| Hoods........... 359 611+ | | | | | |
| Lens mount.. 396 529+ | | | | | |
| **Patent Date** | | **Pat. No.** | | **Patent Title** | **Patentee** |
| 10-Apr-12 | | 8,152,316 | T | Imaging device | Tsuji; Yoshifumi |
| 7-Feb-12 | | 8,111,984 | T | Matte box assembly | Wood; Dennis |
| 8-Nov-11 | | 8,054,545 | T | Lens hood for a camera lens | Cheng; Ming-Chung |
| 8-Feb-11 | | 7,883,222 | T | Lens hood for a camera | Elias; James |
| 12-Oct-10 | | 7,813,639 | T | Camera cover | Yoneji; Osamu ( |
| 1-Dec-09 | | 7,625,140 | T | Light shielding device of a lens | Miya; Kota |
| 9-Sep-08 | | 7,424,214 | T | Push-pull type camera module for portable wireless terminal | Kim; Han-Sik |
| 10-Jun-08 | | 7,386,229 | T | Camera LCD screen hood and viewing device | Schmidt; Louis J. |
| 15-Apr-08 | | 7,359,131 | T | Lens positioning systems and methods | Gutierrez; Roman C. |
| 22-Aug-06 | | 7,093,944 | T | Light shielding structure of a lens barrel | Tanaka; Hitoshi |
| 2-May-06 | | 7,037,007 | T | Lens hood, and imaging device | Ohmori; Norikatsu |
| 17-Jan-06 | | 6,986,584 | T | Arrangement for a rotatable eyepiece cap | Benner; Thomas |
| 6-Dec-05 | | 6,971,754 | T | Lens cap retention arrangement | Flora; Lapthe Chau |
| 5-Jun-01 | | 6,243,540 | T | Lens barrel assembly | Kume; Hideaki |
| 30-Jun-92 | | 5,126,881 | T | Lens hood for a photographic lens | Crema; Rolf |
| 14-Apr-92 | | 5,105,312 | T | Lens mount accessory system | Tiffen; Ira |
| 20-Mar-90 | | 4,909,617 | T | Camera hood with pivoting lens cap | Boyd; Jeffrey M. |
| 9-Sep-86 | | 4,610,517 | T | Clamp for lens hood | Fukino; Kunihiro |
| 6-Aug-85 | | 4,533,212 | T | Accessory holding device for optical instrument | Shimizu; Seiichi |
| 24-May-83 | | 4,384,767 | T | Clamping device for camera accessory or hood | Kawai; Tohru |
| 20-Oct-81 | | 4,295,706 | T | Combined lens cap and sunshade for a camera | Frost; George H. |
| 30-Jan-79 | | 4,137,540 | T | Camera matte box | Curtis; Jack |
| 21-Nov-78 | | 4,126,375 | T | Lens shield means | Bull; David W. |
| 11-Jan-77 | | 4,002,402 | T | Lens hood for zoom lenses | Mito; Hiroshi |
| 30-Sep-75 | | 3,909,107 | T | Hood for the lens of optical instruments with pivotally mounted lens cover | Numbers; Jody L. |
| 8-Oct-74 | | 3,840,883 | T | CAMERA LENS HOOD | Choate; J. Robert |
| 19-Oct-71 | | 3,614,196 | T | COMBINED LENS HOOD AND FILTER SUPPORT | Schlapp; Werner |
| 30-Mar-71 | | 3,572,905 | T | LENS HOOD | Schlapp; Werner |

prior arts, it is also the invaluable source of examining for comparison, specifications and claims for different inventions. The ways they are worded and presented are quite different in even identical subjects.

| Receipt date: 10/22/2013 | Application Number | 14028786 | 14028786 - GAU: 2852 |
|---|---|---|---|
| | Filing Date | 2013-09-17 | |
| **INFORMATION DISCLOSURE STATEMENT BY APPLICANT** ( Not for submission under 37 CFR 1.99) | First Named Inventor | Roy Lique | |
| | Art Unit | 2872 | |
| | Examiner Name | | |
| | Attorney Docket Number | Camera Coup | |

If you wish to add additional U.S. Patent citation information please click the Add button.     Add

**U.S.PATENT APPLICATION PUBLICATIONS**     Remove

| Examiner Initial* | Cite No | Publication Number | Kind Code¹ | Publication Date | Name of Patentee or Applicant of cited Document | Pages,Columns,Lines where Relevant Passages or Relevant Figures Appear |
|---|---|---|---|---|---|---|
| | 1 | | | | | |

If you wish to add additional U.S. Published Application citation information please click the Add button.     Add

**FOREIGN PATENT DOCUMENTS**     Remove

| Examiner Initial* | Cite No | Foreign Document Number³ | Country Code² ᵢ | Kind Code⁴ | Publication Date | Name of Patentee or Applicant of cited Document | Pages,Columns,Lines where Relevant Passages or Relevant Figures Appear | T⁵ |
|---|---|---|---|---|---|---|---|---|
| | 1 | | | | | | | ☐ |

If you wish to add additional Foreign Patent Document citation information please click the Add button     Add

**NON-PATENT LITERATURE DOCUMENTS**     Remove

| Examiner Initials* | Cite No | Include name of the author (in CAPITAL LETTERS), title of the article (when appropriate), title of the item (book, magazine, journal, serial, symposium, catalog, etc), date, pages(s), volume-issue number(s), publisher, city and/or country where published. | T⁵ |
|---|---|---|---|
| | 1 | | ☐ |

If you wish to add additional non-patent literature document citation information please click the Add button     Add

**EXAMINER SIGNATURE**

| Examiner Signature | /Christopher Mahoney/ | Date Considered | 04/03/2015 |
|---|---|---|---|

*EXAMINER: Initial if reference considered, whether or not citation is in conformance with MPEP 609. Draw line through a citation if not in conformance and not considered. Include copy of this form with next communication to applicant.

¹ See Kind Codes of USPTO Patent Documents at www.USPTO.GOV or MPEP 901.04. ² Enter office that issued the document, by the two-letter code (WIPO Standard ST.3). ³ For Japanese patent documents, the indication of the year of the reign of the Emperor must precede the serial number of the patent document. ⁴ Kind of document by the appropriate symbols as indicated on the document under WIPO Standard ST.16 if possible. ⁵ Applicant is to place a check mark here if English language translation is attached.

EFS Web 2.1.17     ALL REFERENCES CONSIDERED EXCEPT WHERE LINED THROUGH. /C.M./

Receipt date: 10/22/2013
14028786 - GAU: 2852

Doc code: IDS
Doc description: Information Disclosure Statement (IDS) Filed

PTO/SB/08a (01-10)
Approved for use through 07/31/2012. OMB 0651-0031
U.S. Patent and Trademark Office, U.S. DEPARTMENT OF COMMERCE
Under the Paperwork Reduction Act of 1995, no persons are required to respond to a collection of information unless it contains a valid OMB control number

| | |
|---|---|
| **INFORMATION DISCLOSURE STATEMENT BY APPLICANT** ( Not for submission under 37 CFR 1.99 ) | |

| Application Number | 14028786 |
|---|---|
| Filing Date | 2013-09-17 |
| First Named Inventor | Roy Lique |
| Art Unit | 2872 |
| Examiner Name | |
| Attorney Docket Number | Camera Coup |

| | | U.S.PATENTS | | | | Remove |
|---|---|---|---|---|---|---|
| Examiner Initial* | Cite No | Patent Number | Kind Code1 | Issue Date | Name of Patentee or Applicant of cited Document | Pages,Columns,Lines where Relevant Passages or Relevant Figures Appear |
| | 1 | 8111984 | | 2012-02-07 | Wood; Dennis | |
| | 2 | 8054545 | | 2011-11-08 | Cheng; Ming-Chung | |
| | 3 | 7813639 | | 2010-10-12 | Yoneji; Osamu | |
| | 4 | 6243540 | | 2001-06-05 | Kume, Hideaki | |
| | 5 | 5126881 | | 1992-06-30 | Crema; Rolf | |
| | 6 | 4533212 | | 1985-08-06 | Shimizu; Seiichi | |
| | 7 | 4384767 | | 1983-05-24 | Kawai; Tohru | |
| | 8 | 4295706 | | 1981-10-20 | Frost; George H. | |

EFS Web 2.1.17    ALL REFERENCES CONSIDERED EXCEPT WHERE LINED THROUGH.  /C.M./

| Receipt date: 10/22/2013 | | Application Number | 14028786 | 14028786 - GAU: 2852 |
|---|---|---|---|---|
| **INFORMATION DISCLOSURE STATEMENT BY APPLICANT** ( Not for submission under 37 CFR 1.99) | | Filing Date | 2013-09-17 | |
| | | First Named Inventor | Roy Lique | |
| | | Art Unit | 2872 | |
| | | Examiner Name | | |
| | | Attorney Docket Number | Camera Coup | |

| | | | | |
|---|---|---|---|---|
| 9 | 4137540 | 1979-01-30 | Curtis; Jack | |
| 10 | 3909107 | 1975-09-30 | Numbers; Jody L. | |
| 11 | 3840883 | 1974-10-08 | Choate; J. Robert | |
| 12 | 3614196 | 1971-10-19 | Schlapp; Werner | |
| 13 | 8045277 | 2011-10-25 | Iwasaki; Tetsuya | |
| 14 | 7386229 | 2008-06-10 | Schmidt; Louis J. | |
| 15 | 7359131 | 2008-04-15 | Gutierrez; Roman C. | |
| 16 | 7161749 | 2007-01-09 | Sakurai; Nobumasa | |
| 17 | 7031081 | 2006-04-18 | Petroff; Robert | |
| 18 | 6243540 | 2001-06-05 | Kume; Hideaki | |
| 19 | 7982981 | 2011-07-19 | Fukino; Kunihiro | |

EFS Web 2 1.17    ALL REFERENCES CONSIDERED EXCEPT WHERE LINED THROUGH. /C.M.

## Drawings

## Standards for Drawings

(1) Drawings. There are two acceptable categories for presenting drawings in utility and design patent applications:

(a) Black ink. Black and white drawings are normally required. India ink, or its equivalent that secures solid black lines, must be used for drawings, or

(b) Color. On rare occasions, color drawings may be necessary as the only practical medium by which to disclose the subject matter sought to be patented in a utility or design patent application or the subject matter of a statutory invention registration. The color drawings must be of sufficient quality such that all details in the drawings are reproducible in black and white in the printed patent. Color drawings are not permitted in international applications (see PCT Rule 11.13), or in an application, or copy thereof, submitted under the Office electronic filing system.

The Office will accept color drawings in utility or design patent applications and statutory invention registrations only after granting a petition filed under this paragraph explaining why the color drawings are necessary. Any such petition must include the following:

(I) The fee set forth in § 1.17(h);

(ii) Three sets of color drawings; and

(iii) An amendment to the specification to insert (unless the specification contains or has been previously amended to contain) the following language as the first paragraph of the brief description of the drawings:

The patent or application file contains at least one drawing executed in color. Copies of this patent or patent application publication with color drawing(s) will be provided by the Office upon request and payment of the necessary fee.

(2) Photographs

(a) Black and white. Photographs, including photocopies of photographs, are not ordinarily permitted in utility and design patent applications. The Office will accept photographs in utility and design patent applications, however, if photographs are the only practicable medium for illustrating the claimed invention. For example, photographs or photomicrographs of electrophoresis gels, blots (e.g., immuno-logical, western, southern, and northern), autoradiographs, cell cultures (stained and unstained), histological tissue cross sections (stained and unstained), animals, plants, in vivo imaging, thin-layer chromatography plates, crystalline structures, and, in a design patent application, ornamental effects, are acceptable. If the subject matter of the application admits of illustration by a drawing, the examiner may require a drawing in place of the photograph. The photographs must be of sufficient quality so that all details in the photographs are reproducible in the printed patent.

(b)Color photographs. Color photographs will be accepted in utility and design patent applications if the conditions for accepting color drawings and black and white photographs have been satisfied. See paragraphs (a)(2) and (b)(1) of this section.

(3) Identification of drawings - Identifying indicia should be provided, and if provided, should include the title of the invention, inventor's name, and application number, or docket number (if any) if an application number has not been assigned to the application. If this information is provided, it must be placed on the front of each sheet within the top margin. Each drawing sheet submitted after the filing date of an application must be identified as either "Replacement Sheet" or "New Sheet" pursuant to § 1.121(d). If a marked-up copy of any amended drawing figure including annotations indicating the changes made is filed, such marked-up copy must be clearly labeled as "Annotated Sheet" pursuant to § 1.121(d)(1).

(4) Graphic forms in drawings - Chemical or mathematical formulae, tables, and waveforms may be submitted as drawings and are subject to the same requirements as drawings. Each chemical or mathematical formula must be labeled as a separate figure, using brackets when necessary, to show that information is properly integrated. Each group of waveforms must be presented as a single figure, using a common vertical axis with time extending along the horizontal axis. Each individual waveform discussed in the specification must be identified with a separate letter designation adjacent to the vertical axis.

(5) Margins - The sheets must not contain frames around the sight (i.e., the usable surface), but should have scan target points (i.e., cross-hairs) printed on two cattycorner margin corners. Each sheet must include a top margin of at least 2.5 cm (1 inch), a left side margin of at least 2.5 cm (1 inch), a right side margin of at least 1.5 cm (5/8 inch), and a bottom margin of at least 1 cm (3/8 inch), thereby leaving a sight no greater than 17 cm by 26.2 cm on 21 cm by 29.7 cm (DIN size A4) drawing sheets, and a sight no greater than 17.6 cm by 24.4 cm (6 15/16 by 9 5/8 inches) on 21.6 cm by 27.9 cm (8 1/2 by 11 inch) drawing sheets.

(6) Views - The drawing must contain as many views as necessary to show the invention. The views may be plan, elevation, section, or perspective views. Detail views of portions of elements, on a larger scale if necessary, may also be used. All views of the drawing must be grouped together and arranged on the sheet(s) without wasting space, preferably in an upright position, clearly separated from one another, and must not be included in the sheets containing the specifications, claims, or abstract. Views must not be connected by projection lines and must not contain center lines. Waveforms of electrical signals may be connected by dashed lines to show the relative timing of the waveforms.

(a) Exploded views – Exploded views with the separated parts embraced by a bracket, to show the relationship or order of assembly of various parts are permissible. When an exploded view is shown in a figure that is on the same sheet as another figure, the exploded view should be placed in brackets.

(b) Partial views - When necessary, a view of a large machine or device in its entirety may be broken into partial views on a single sheet, or extended over several sheets if there is no loss in facility of understanding the view. Partial views drawn on separate sheets must always be capable of being linked edge to edge so that no partial view contains parts of another partial view. A smaller scale view should be included showing the whole formed by the partial views and indicating the positions of the parts shown. When a portion of a view is enlarged for magnification purposes, the view and the enlarged view must each be labeled as separate views.

(i) Where views on two or more sheets form, in effect, a single complete view, the views on the several sheets must be so arranged that the complete figure can be assembled without concealing any part of any of the views appearing on the various sheets.

(ii) A very long view may be divided into several parts placed one above the other on a single sheet. However, the relationship between the different parts must be clear and unambiguous.

(c) Sectional views. The plane upon which a sectional view is taken should be indicated on the view from which the section is cut by a broken line. The ends of the broken line should be designated by Arabic or Roman numerals corresponding to the view number of the sectional view, and should have arrows to indicate the direction of sight. Hatching must be used to indicate section portions of an object, and must be made by regularly spaced oblique parallel lines spaced sufficiently apart to enable the lines to be distinguished without difficulty. Hatching should not impede the clear reading of the reference characters and lead lines. If it is not possible to place reference characters outside the hatched area, the hatching may be broken off wherever reference characters are inserted. Hatching must be at a substantial angle to the surrounding axes or principal lines, preferably 45 degrees. A cross section must be set out and drawn to show all of the materials as they are shown in the view from which the cross section was taken. The parts in cross section must show proper material(s) by hatching with regularly spaced parallel oblique strokes, the space between strokes being chosen on the basis of the total area to be hatched. The various parts of a cross section of the same item should be hatched in the same manner and should accurately and graphically indicate the nature of the material(s) that is illustrated in cross section. The hatching of juxtaposed different elements must be angled in a different way. In the case of large areas, hatching may be confined to an edging drawn around the entire inside of the outline of the area to be hatched. Different types of hatching should have different conventional meanings as regards the nature of a material seen in cross section.

(d) Alternate position. A moved position may be shown by a broken line superimposed upon a suitable view if this can be done without crowding; otherwise, a separate view must be used for this purpose.

(e) Modified forms. Modified forms of construction must be shown in separate views.

(7) Arrangement of views - One view must not be placed upon another or within the outline of another. All views on the same sheet should stand in the same direction and, if possible, stand so that they can be read with the sheet held in an upright position. If views wider than the width of the sheet are necessary for the clearest illustration of the invention, the sheet may be turned on its side so that the top of the sheet, with the appropriate top margin to be used as the heading space, is on the right-hand side. Words must appear in a horizontal, left-to-right fashion when the page is either upright or turned so that the top becomes the right side, except for graphs utilizing standard scientific convention to denote the axis of abscissas (of X) and the axis of ordinates (of Y).

(8) Front page view - The drawing must contain as many views as necessary to show the invention. One of the views should be suitable for inclusion on the front page of the patent application publication and patent as the illustration of the invention. Views must not be connected by projection lines and must not contain center lines. Applicant may suggest a single view (by figure number) for inclusion on the front page of the patent application publication and patent.

(9) Scale - The scale to which a drawing is made must be large enough to show the mechanism without crowding when the drawing is reduced in size to two-thirds in reproduction. Indications such as "actual size" or "scale 1/2" on the drawings are not permitted since these lose their meaning with reproduction in a different format.

(10) Character of lines, numbers, and letters - All drawings must be made by a process that will give them satisfactory reproduction characteristics. Every line, number, and letter must be durable, clean, black (except for color drawings), sufficiently dense and dark, and uniformly thick and well defined. The weight of all lines and letters must be heavy enough to permit adequate reproduction. This requirement applies to all lines however fine, to shading, and to lines representing cut surfaces in sectional views. Lines and strokes of different thicknesses may be used in the same drawing where different thicknesses have a different meaning.

(11) Shading - The use of shading in views is encouraged if it aids in understanding the invention and if it does not reduce legibility. Shading is used to indicate the surface or shape of spherical, cylindrical, and conical elements of an object. Flat parts may also be lightly shaded. Such shading is preferred in the case of parts shown in perspective, but not for cross sections. See paragraph (h)(3) of this section. Spaced lines for shading are preferred. These lines must be thin, as few in number as practicable, and they must contrast with the rest of the drawings. As a substitute for shading, heavy lines on the shade side of objects can be used

except where they superimpose on each other or obscure reference characters. Light should come from the upper left corner at an angle of 45 degrees. Surface delineations should preferably be shown by proper shading. Solid black shading areas are not permitted, except when used to represent bar graphs or color.

(12) Symbols - Graphical drawing symbols may be used for conventional elements when appropriate. The elements for which such symbols and labeled representations are used must be adequately identified in the specification. Known devices should be illustrated by symbols that have a universally recognized conventional meaning and are generally accepted in the art. Other symbols which are not universally recognized may be used, subject to approval by the Office, if they are not likely to be confused with existing conventional symbols, and if they are readily identifiable.

(13) Legends - Suitable descriptive legends may be used subject to approval by the Office, or may be required by the examiner where necessary for understanding of the drawing. They should contain as few words as possible.

(14)Numbers, letters, and reference characters

(a) Reference characters (numerals are preferred), sheet numbers, and view numbers must be plain and legible, and must not be used in association with brackets or inverted commas, or enclosed within outlines, e.g., encircled. They must be oriented in the same direction as the view so as to avoid having to rotate the sheet. Reference characters should be arranged to follow the profile of the object depicted.

(b) The English alphabet must be used for letters, except where another alphabet is customarily used, such as the Greek alphabet to indicate angles, wavelengths, and mathematical formulas.

(c) Numbers, letters, and reference characters must measure at least .32 cm (1/8 inch) in height. They should not be placed in the drawing so as to interfere with its comprehension. Therefore, they should not cross or mingle with the lines. They should not be placed upon hatched or shaded surfaces. When necessary, such as indicating a surface or cross section, a reference character may be underlined and a blank space may be left in the hatching or shading where the character occurs so that it appears distinct.

(d) The same part of an invention appearing in more than one view of the drawing must always be designated by the same reference character, and the same reference character must never be used to designate different parts.

(e) Reference characters not mentioned in the description shall not appear in the drawings. Reference characters mentioned in the description must appear in the drawings.

(15) Lead lines - Lead lines are those lines between the reference characters and the details referred to. Such lines may be straight or curved and should be as short as possible. They must originate in the immediate proximity of the reference character and extend to the feature indicated. Lead lines must not cross each other. Lead lines are required for each reference character except for those which indicate the surface or cross section on which they are placed. Such a reference character must be underlined to make it clear that a lead line has not been left out by mistake. Lead lines must be executed in the same way as lines in the drawing. See paragraph (1) of this section.

(16) Arrows - Arrows may be used at the ends of lines, provided that their meaning is clear, as follows:

(a) On a lead line, a freestanding arrow to indicate the entire section towards which it points;

(b) On a lead line, an arrow touching a line to indicate the surface shown by the line looking along the direction of the arrow

(c) To show the direction of movement.

(17) Copyright or Mask Work Notice - A copyright or mask work notice may appear in the drawing, but must be placed within the sight of the drawing immediately below the figure representing the copyright or mask work material and be limited to letters having a print size of .32 cm. to .64 cm. (1/8 to 1/4 inches) high. The content of the notice must be limited to only those elements provided for by law. For example, "©1983 John Doe" (17 U.S.C. 401) and "*M* John Doe" (17 U.S.C. 909) would be properly limited and, under current statutes, legally sufficient notices of copyright and mask work, respectively. Inclusion of a copyright or mask work notice will be permitted only if the authorization language set forth in 1.71(e) is included at the beginning (preferably as the first paragraph) of the specification.

(18)Numbering of sheets of drawings - The sheets of drawings should be numbered in consecutive Arabic numerals, starting with 1, within the sight as defined in paragraph (5) of this section. These numbers, if present, must be placed in the middle of the top of the sheet, but not in the margin. The numbers can be placed on the right-hand side if the drawing extends too close to the middle of the top edge of the usable surface. The drawing sheet numbering must be clear and larger than the numbers used as reference characters to avoid confusion. The number of each sheet should be shown by two Arabic numerals placed on either side of an oblique line, with the first being the sheet number and the second being the total number of sheets of drawings, with no other marking.

(19) Numbering of views

(a) The different views must be numbered in consecutive Arabic numerals, starting with 1, independent of the numbering of the sheets and, if possible, in the order in which they appear on the drawing sheet(s). Partial views intended to form one complete view, on one or several sheets must be identified by the same number followed by a capital letter. View numbers must be preceded by the abbreviation "FIG." Where only a single view is used in an application to illustrate the claimed invention, it must not be numbered and the abbreviation "FIG." must not appear.

(b) Numbers and letters identifying the views must be simple and clear and must not be used in association with brackets, circles, or inverted commas. The view numbers must be larger than the numbers used for reference characters.

(20) Security markings - Authorized security markings may be placed on the drawings provided they are outside the sight, preferably centered in the top margin.

(21) Corrections - Any corrections on drawings submitted to the Office must be durable and permanent.

(22) Holes - No holes should be made by applicant in the drawing sheets.

(23) Types of drawings - See § 1.152 for design drawings, § 1.165 for plant drawings, and § 1.173(a)(2) for reissue drawings

## How I prepared the drawings

If there was any aspect that could have easily derailed my application for a patent, it was the preparation of drawings. Having gone through a cursory reading of all the provisions pertaining to filing an application for a patent, I concluded that it was better to start thinking about drawings while the object of the invention was being prototyped.

As simple as the invention was, the drawing aspect as I realized, turned out to be very demanding especially when half a dozen views were to be presented. I had some drawing templates for previous different purposes; I was willing to buy more, but nothing seemed to make the job easier. Drawing was a dilemma I had to endure for some days threatening to derail all my plans pertaining to the invention.

Having a draftsman do the drawings for me was an option I was hesitantly willing to take, but the services are expensive and needed to be repeated as often as necessary to accommodate changes in the prototype, thereby multiplying the cost. I have friends who are proficient with auto cad and may be one or two friendly requests were all right with them. But with a number of views required and each view vulnerable to changes as the prototyping went on, friendship was doomed to be strained because drawings require a considerable length of time to do.

Finally, I reviewed the features of an old drawing software that I have for a different purpose. To my surprise, the software has features that I was not aware of before, which allowed me to do the following in the adjoining image:

1. Import picture A into the work area
2. Trace through the contours of picture A that are needed in picture B
3. Delete picture A leaving only its outline that will make up picture B

4. Smooth and break some lines if necessary
5. Add shapes if necessary
6. Move drawing elements closer or farther away if necessary
7. Delete some elements if necessary

In half an hour I had a graphic presentation of the invention in the form of Picture B. Using the Windows snipping tool while still in the program, I saved the standard portion of the image in a file that could be altered, added with shapes and symbols, and be a source of different views, thus avoiding the retracing of the image again for later presentations. When it became necessary to present different views of the invention, all I needed was add few shapes and symbols to the master image.

In the adjoining image, the camera mount assembly and the master cylinder are standard images to which I only added the arrow symbols, the numerals, the boxed images, and others, to derive different views of the invention.

I am certain that there is plenty of drawing software that would allow one to draw graphics similar to what I have done. Many of them are free for the taking. In the application for a patent, drawing then should not be a fearsome phase, I now conclude.

## Drawings

Drawing - Figure 1

Fig.1 shows a lens guard with slip type camera mount assembly and a coupler ring positioned between a camera and a sighting device

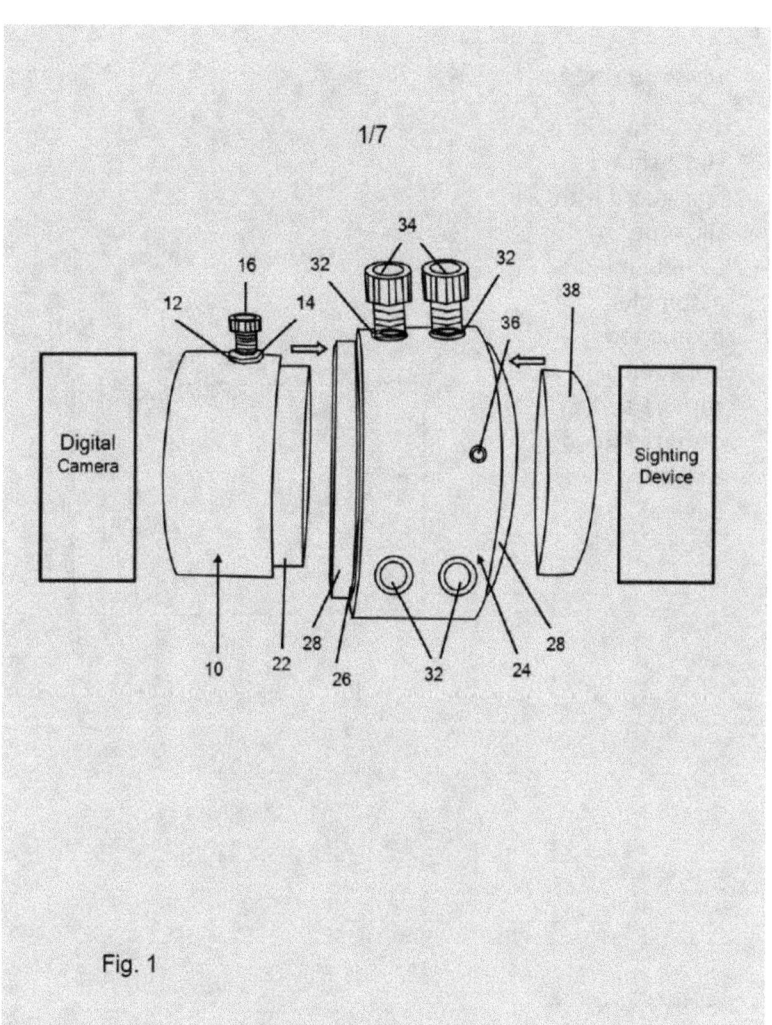

Fig. 1

Drawing - Figure 2

Fig. 2 shows a lens guard with slip type camera mount assembly, positioned between a camera and a camera filter or a sighting device.

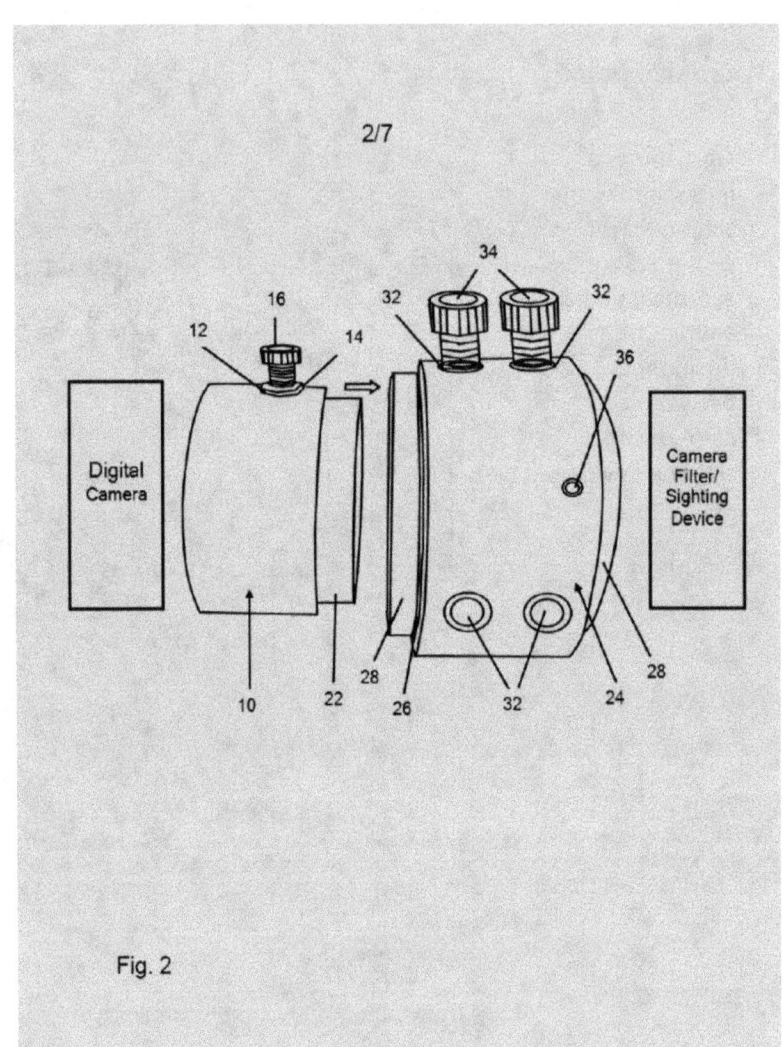

Fig. 2

Drawing - Figure 3

Fig. 3 shows a lens guard with screw type camera mount assembly and a coupler ring, positioned between a camera and a sighting device.

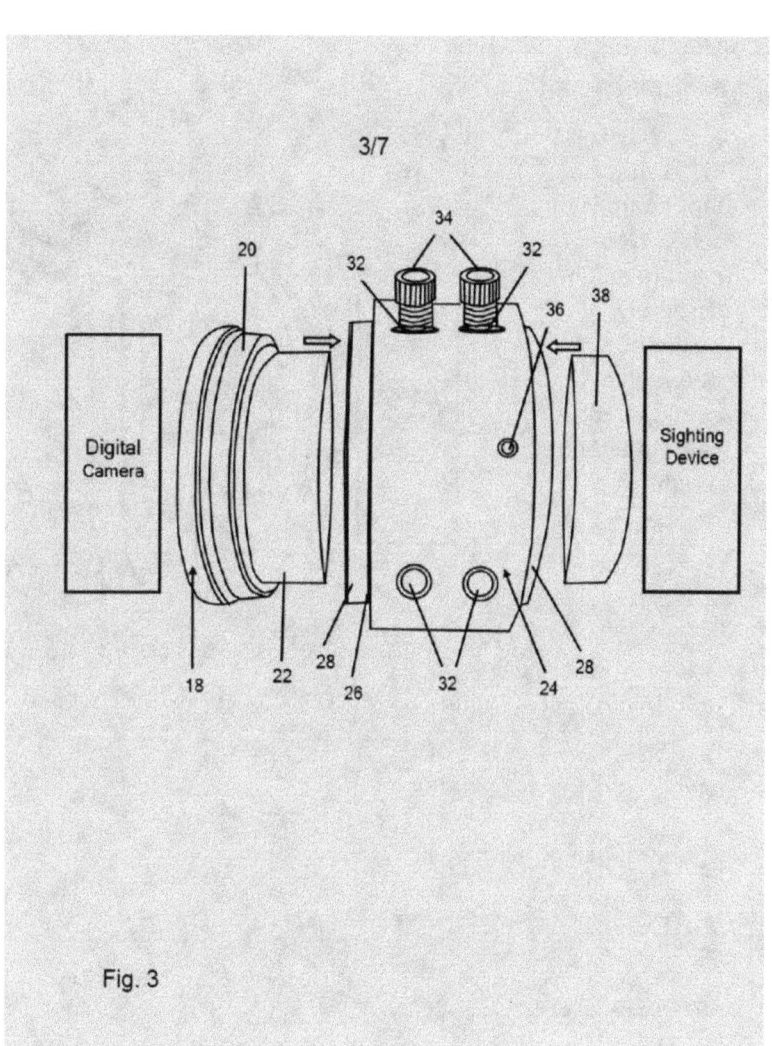

Fig. 3

Drawing - Figure 4

Fig. 4 shows a lens guard with screw type camera mount assembly, positioned between a camera and a camera filter or a sighting device.

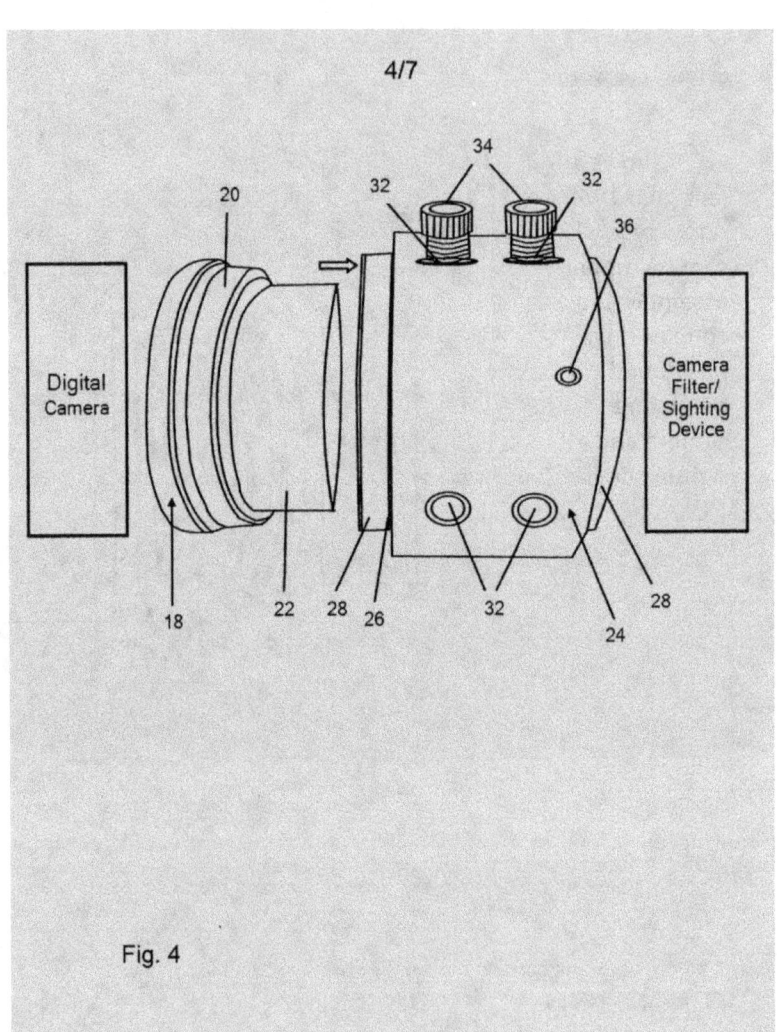

Fig. 4

Drawing - Figure 5

Fig. 5 shows a slip type barrel track directly securing a camera filter or a sighting device, positioned between a camera and a sighting device.

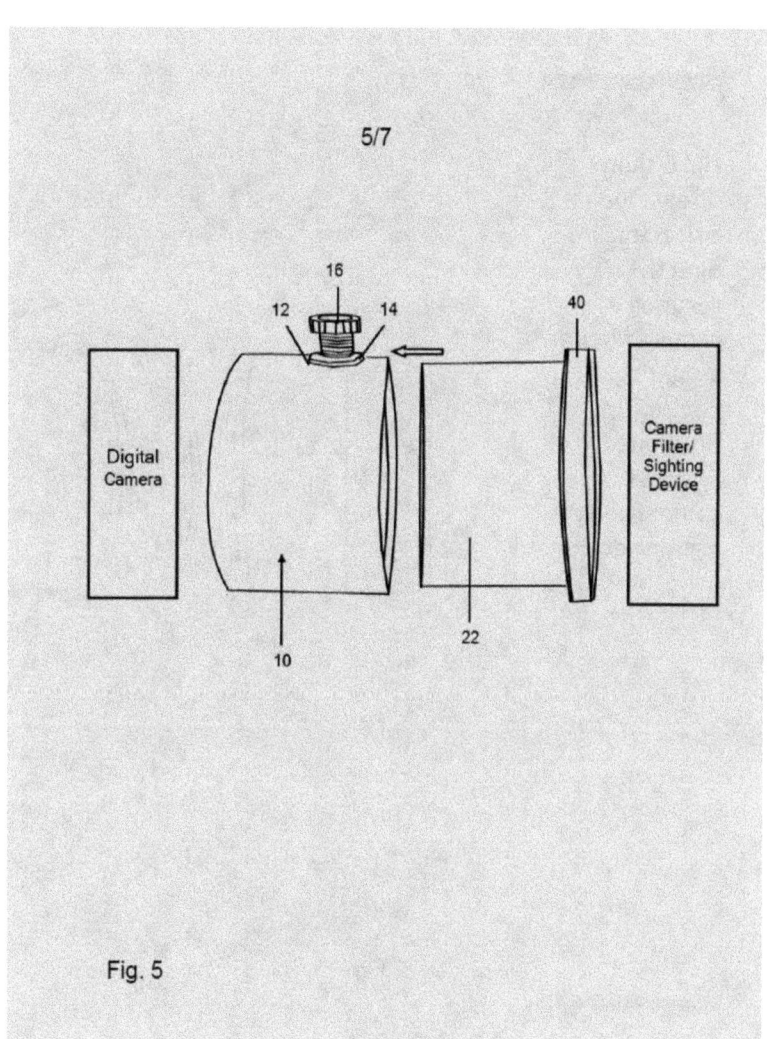

Fig. 5

Drawing - Figure 6

Fig. 6 shows a screw type barrel track directly securing a camera filter or a sighting device, positioned between a camera and a sighting device.

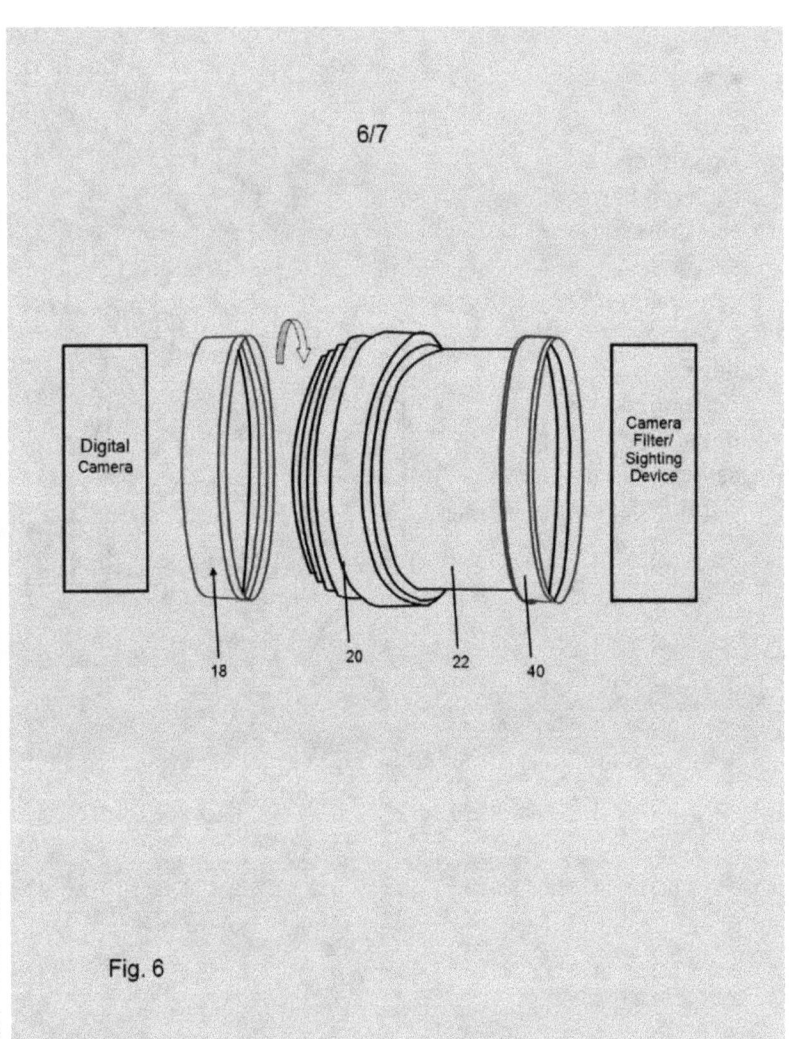

6/7

Digital Camera

Camera Filter/ Sighting Device

18

20

22

40

Fig. 6

Drawing - Figure 7

Fig. 7 is an end
view of a
primary ring
with a stack of
outer and inner
adapter rings.

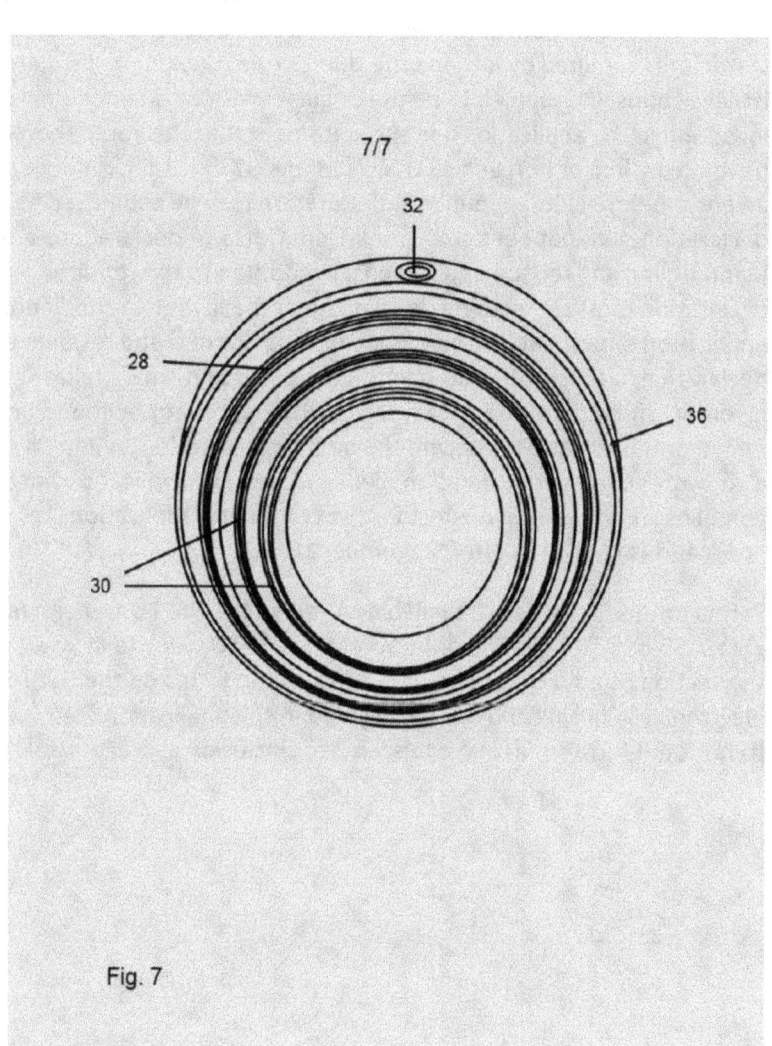

Fig. 7

## Oath or Declaration, Signature

An oath or declaration is a formal statement that must be made by the inventor in a non-provisional application. Each inventor must sign an oath or declaration that includes certain statements required by law and the USPTO rules, including the statement that he or she believes himself or herself to be the original inventor or an original joint inventor of a claimed invention in the application and the statement that the application was made or authorized to be made by him or her. See 35 U.S.C 115 and 37 CFR 1.63. An oath must be sworn to by the inventor before a notary public. A declaration may be submitted in lieu of an oath. A declaration does not need to be notarized. Oaths or declarations are required for design, plant, utility, and reissue applications. In addition to the required statements, the oath or declaration must set forth the legal name of the inventor, and, if not provided in an application data sheet, the inventor's mailing address and residence. In lieu of an oath or declaration, a substitute statement may be signed by the applicant with respect to an inventor who is deceased, legally incapacitated, cannot be found or reached after diligent effort, or has refused to execute the oath or declaration. When filing a continuing application, a copy of the oath or declaration filed in the earlier application may be used provided that it complies with the rules in effect for the continuing application (i.e., the rules that apply to applications filed on or after September 16, 2012).

Forms for declarations are available by calling the USPTO General Information Services at 800-786-9199 or 571-272-1000 or by accessing USPTO website at www.uspto.gov, indexed under the section titled "Forms, Patents." Most of the forms on the USPTO website are electronically fillable and can be included in the application filed via EFS-Web without having to print the form out in order to scan it for inclusion as a PDF attachment to the application.

## Declaration – Page 1

Degree of difficulty of filling highlighted items: EASY

PTO/SB/01A (09-12)
Approved for use through 01/31/2014. OMB 0651-0032
U.S. Patent and Trademark Office; U.S. DEPARTMENT OF COMMERCE
Under the Paperwork Reduction Act of 1995, no persons are required to respond to a collection of information unless it displays a valid OMB control number.

**DECLARATION (37 CFR 1.63) FOR UTILITY OR DESIGN APPLICATION USING AN APPLICATION DATA SHEET (37 CFR 1.76)**

| | |
|---|---|
| **Title of Invention** | Digital Camera Lens Guard and Use Extender |

As the below named inventor(s), I/we declare that:

This declaration is directed to:
- ■ The attached application, or
- ☐ United States application or PCT international application number ____________ filed on ____________
- ☐ As amended on ____________ (if applicable);

I/we believe that I/we am/are the original and first inventor(s) of the subject matter which is claimed and for which a patent is sought;

I/we have reviewed and understand the contents of the above-identified application, including the claims, as amended by any amendment specifically referred to above;

I/we acknowledge the duty to disclose to the United States Patent and Trademark Office all information known to me/us to be material to patentability as defined in 37 CFR 1.56, including for continuation-in-part applications, material information which became available between the filing date of the prior application and the national or PCT International filing date of the continuation-in-part application. The above-identified application was made or authorized to be made by me/us.

**WARNING:**

Petitioner/applicant is cautioned to avoid submitting personal information in documents filed in a patent application that may contribute to identity theft. Personal information such as social security numbers, bank account numbers, or credit card numbers (other than a check or credit card authorization form PTO-2038 submitted for payment purposes) is never required by the USPTO to support a petition or an application. If this type of personal information is included in documents submitted to the USPTO, petitioners/applicants should consider redacting such personal information from the documents before submitting them to the USPTO. Petitioner/applicant is advised that the record of a patent application is available to the public after publication of the application (unless a non-publication request in compliance with 37 CFR 1.213(a) is made in the application) or issuance of a patent. Furthermore, the record from an abandoned application may also be available to the public if the application is referenced in a published application or an issued patent (see 37 CFR 1.14). Checks and credit card authorization forms PTO-2038 submitted for payment purposes are not retained in the application file and therefore are not publicly available.

All statements made herein of my/our own knowledge are true, all statements made herein on information and belief are believed to be true, and further that these statements were made with the knowledge that willful false statements and the like are punishable by fine or imprisonment, or both, under 18 U.S.C. 1001, and may jeopardize the validity of the application or any patent issuing thereon. I hereby acknowledge that any willful false statement made in this declaration is punishable under 18 U.S.C. 1001 by fine or imprisonment of not more than five (5), or both.

FULL NAME OF INVENTOR(S)

Inventor one: Roy Lique Date: 2013-09-15
Signature: /Roy Lique/ Citizen of: United States

Inventor two: ____________ Date: ____________
Signature: ____________ Citizen of: ____________

☐ Additional inventors or a legal representative are being named on ____________ additional form(s) attached hereto.

This collection of information is required by 35 U.S.C. 115 and 37 CFR 1.63. The information is required to obtain or retain a benefit by the public which is to file (and by the USPTO to process) an application. Confidentiality is governed by 35 U.S.C. 122 and 37 CFR 1.11 and 1.14. This collection is estimated to take 1 minute to complete, including gathering, preparing, and submitting the completed application form to the USPTO. Time will vary depending upon the individual case. Any comments on the amount of time you require to complete this form and/or suggestions for reducing this burden, should be sent to the Chief Information Officer, U.S. Patent and Trademark Office, U.S. Department of Commerce, P.O. Box 1450, Alexandria, VA 22313-1450. DO NOT SEND FEES OR COMPLETED FORMS TO THIS ADDRESS. **SEND TO: Commissioner for Patents, P.O. Box 1450, Alexandria, VA 22313-1450.**

*If you need assistance in completing the form, call 1-800-PTO-9199 and select option 2.*

**Declaration – Page 2**

Privacy
Statement

### Privacy Act Statement

The **Privacy Act of 1974 (P.L. 93-579)** requires that you be given certain information in connection with your submission of the attached form related to a patent application or patent. Accordingly, pursuant to the requirements of the Act, please be advised that: (1) the general authority for the collection of this information is 35 U.S.C. 2(b)(2); (2) furnishing of the information solicited is voluntary; and (3) the principal purpose for which the information is used by the U.S. Patent and Trademark Office is to process and/or examine your submission related to a patent application or patent. If you do not furnish the requested information, the U.S. Patent and Trademark Office may not be able to process and/or examine your submission, which may result in termination of proceedings or abandonment of the application or expiration of the patent.

The information provided by you in this form will be subject to the following routine uses:

1. The information on this form will be treated confidentially to the extent allowed under the Freedom of Information Act (5 U.S.C. 552) and the Privacy Act (5 U.S.C 552a). Records from this system of records may be disclosed to the Department of Justice to determine whether disclosure of these records is required by the Freedom of Information Act.
2. A record from this system of records may be disclosed, as a routine use, in the course of presenting evidence to a court, magistrate, or administrative tribunal, including disclosures to opposing counsel in the course of settlement negotiations.
3. A record in this system of records may be disclosed, as a routine use, to a Member of Congress submitting a request involving an individual, to whom the record pertains, when the individual has requested assistance from the Member with respect to the subject matter of the record.
4. A record in this system of records may be disclosed, as a routine use, to a contractor of the Agency having need for the information in order to perform a contract. Recipients of information shall be required to comply with the requirements of the Privacy Act of 1974, as amended, pursuant to 5 U.S.C. 552a(m).
5. A record related to an International Application filed under the Patent Cooperation Treaty in this system of records may be disclosed, as a routine use, to the International Bureau of the World Intellectual Property Organization, pursuant to the Patent Cooperation Treaty.
6. A record in this system of records may be disclosed, as a routine use, to another federal agency for purposes of National Security review (35 U.S.C. 181) and for review pursuant to the Atomic Energy Act (42 U.S.C. 218(c)).
7. A record from this system of records may be disclosed, as a routine use, to the Administrator, General Services, or his/her designee, during an inspection of records conducted by GSA as part of that agency's responsibility to recommend improvements in records management practices and programs, under authority of 44 U.S.C. 2904 and 2906. Such disclosure shall be made in accordance with the GSA regulations governing inspection of records for this purpose, and any other relevant (i.e., GSA or Commerce) directive. Such disclosure shall not be used to make determinations about individuals.
8. A record from this system of records may be disclosed, as a routine use, to the public after either publication of the application pursuant to 35 U.S.C. 122(b) or issuance of a patent pursuant to 35 U.S.C. 151. Further, a record may be disclosed, subject to the limitations of 37 CFR 1.14, as a routine use, to the public if the record was filed in an application which became abandoned or in which the proceedings were terminated and which application is referenced by either a published application, an application open to public inspection or an issued patent.
9. A record from this system of records may be disclosed, as a routine use, to a Federal, State, or local law enforcement agency, if the USPTO becomes aware of a violation or potential violation of law or regulation.

## Certificate of Micro Entity Status

- (a) To establish micro entity status under this paragraph, the applicant must certify that:

  - (1) The applicant qualifies as a small entity as defined in § 1.27;

  - (2) Neither the applicant nor the inventor nor a joint inventor has been named as the inventor or a joint inventor on more than four previously filed patent applications, other than applications filed in another country, provisional applications under **35 U.S.C. 111(b)**, or international applications for which the basic national fee under **35 U.S.C. 41(a)** was not paid;

  - (3) Neither the applicant nor the inventor nor a joint inventor, in the calendar year preceding the calendar year in which the applicable fee is being paid, had a gross income, as defined in section 61(a) of the Internal Revenue Code of 1986 (26 U.S.C. 61(a)), exceeding three times the median household income for that preceding calendar year, as most recently reported by the Bureau of the Census; and

  - (4) Neither the applicant nor the inventor nor a joint inventor has assigned, granted, or conveyed, nor is under an obligation by contract or law to assign, grant, or convey, a license or other ownership interest in the application concerned to an entity that, in the calendar year preceding the calendar year in which the applicable fee is being paid, had a gross income, as defined in section 61(a) of the Internal Revenue Code of 1986, exceeding three times the median household income for that preceding calendar year, as most recently reported by the Bureau of the Census.

- (b) An applicant, inventor, or joint inventor is not considered to be named on a previously filed application for purposes of paragraph (a)(2) of this section if the applicant, inventor, or joint inventor has assigned, or is under an obligation by contract or law to assign, all ownership rights in the application as the result of the applicant's, inventor's, or joint inventor's previous employment.

- (c) If an applicant's, inventor's, joint inventor's, or entity's gross income in the preceding calendar year is not in United States dollars, the average currency exchange rate, as reported by the Internal Revenue Service, during that calendar year shall be used to determine whether the applicant's, inventor's, joint inventor's, or entity's gross income exceeds the threshold specified in paragraph (a)(3) or (4) of this section.

- (d) To establish micro entity status under this paragraph, the applicant must certify that:
  - (1) The applicant qualifies as a small entity as defined in **§ 1.27**; and
  - (2)
    - (i) The applicant's employer, from which the applicant obtains the majority of the applicant's income, is an institution of higher education as defined in section 101(a) of the Higher Education Act of 1965 (20 U.S.C. 1001(a)); or
    - (ii) The applicant has assigned, granted, conveyed, or is under an obligation by contract or law, to assign, grant, or convey, a license or other ownership interest in the particular application to such an institution of higher education.

- (e) Micro entity status is established in an application by filing a micro entity certification in writing complying with the requirements of either paragraph (a) or (d) of this section and signed either in compliance with § **1.33(b)**, in an international application filed in a Receiving Office other than the United States Receiving Office by a person authorized to represent the applicant under § 1.455, or in an international design application by a person authorized to represent the applicant under § 1.1041 before the International Bureau where the micro entity certification is filed with the International Bureau. Status as a micro entity must be specifically established in each related, continuing and reissue application in which status is appropriate and desired. Status as a micro entity in one application or patent does not affect the status of any other application or patent, regardless of the relationship of the applications or patents. The refiling of an application under § **1.53** as a continuation, divisional, or continuation-in-part application (including a continued prosecution application under § **1.53(d)** ), or the filing of a reissue application, requires a new certification of entitlement to micro entity status for the continuing or reissue application.

- (f) A fee may be paid in the micro entity amount only if it is submitted with, or subsequent to, the submission of a certification of entitlement to micro entity status.

- (g) A certification of entitlement to micro entity status need only be filed once in an application or patent. Micro entity status, once established, remains in effect until changed pursuant to paragraph (i) of this section. However, a fee may be paid in the micro entity amount only if status as a micro entity as defined in paragraph (a) or (d) of this section is appropriate on the date the fee is being paid. Where an assignment of rights or an obligation to assign rights to other parties who are micro entities occurs subsequent to the filing of a certification of entitlement to micro entity status, a second certification of entitlement to micro entity status is not required.

- (h) Prior to submitting a certification of entitlement to micro entity status in an application, including a related, continuing, or reissue application, a determination of such entitlement should be made pursuant to the requirements of this section. It should be determined that each applicant qualifies for micro entity status under paragraph (a) or (d) of this section, and that any other party holding rights in the invention qualifies for small entity status under **§ 1.27**. The Office will generally not question certification of entitlement to micro entity status that is made in accordance with the requirements of this section.

- (i) Notification of a loss of entitlement to micro entity status must be filed in the application or patent prior to paying, or at the time of paying, any fee after the date on which status as a micro entity as defined in paragraph (a) or (d) of this section is no longer appropriate. The notification that micro entity status is no longer appropriate must be signed by a party identified in § **1.33(b).** Payment of a fee in other than the micro entity amount is not sufficient notification that micro entity status is no longer appropriate. A notification that micro entity status is no longer appropriate will not be treated as a notification that small entity status is also no longer appropriate unless it also contains a notification of loss of entitlement to small entity status under § 1.27(f)(2)[**1.27(g)(2)** ]. Once a notification of a loss of entitlement to micro entity status is filed in the application or patent, a new certification of entitlement to micro entity status is required to again obtain micro entity status.

- (j) Any attempt to fraudulently establish status as a micro entity, or pay fees as a micro entity, shall be considered as a fraud practiced or attempted on the Office. Improperly, and with intent to deceive, establishing status as a micro entity, or paying fees as a micro entity, shall be considered as a fraud practiced or attempted on the Office.

- (k) If status as a micro entity is established in good faith in an application or patent, and fees as a micro entity are paid in good faith in the application or patent, and it is later discovered that such micro entity status either was established in error, or that the Office was not notified of a loss of entitlement to micro entity status as required by paragraph (i) of this section through error, the error will be excused upon compliance with the separate submission and itemization requirements of paragraph (k)(1) of this section and the deficiency payment requirement of paragraph (k)(2) of this section.

- (1) Any paper submitted under this paragraph must be limited to the deficiency payment (all fees paid in error) required for a single application or patent. Where more than one application or patent is involved, separate submissions of deficiency payments are required for each application or patent (see § **1.4(b)** ). The paper must contain an itemization of the total deficiency payment for the single application or patent and include the following information:

  - (i) Each particular type of fee that was erroneously paid as a micro entity, (*e.g.,* basic statutory filing fee, two-month extension of time fee) along with the current fee amount for a small or non-small entity, as applicable;
  - (ii) The micro entity fee actually paid, and the date on which it was paid;
  - (iii) The deficiency owed amount (for each fee erroneously paid); and
  - (iv) The total deficiency payment owed, which is the sum or total of the individual deficiency owed amounts as set forth in paragraph (k)(2) of this section.

- (2) The deficiency owed, resulting from the previous erroneous payment of micro entity fees, must be paid. The deficiency owed for each previous fee erroneously paid as a micro entity is the difference between the current fee amount for a small entity or non-small entity, as applicable, on the date the deficiency is paid in full and the amount of the previous erroneous micro entity fee payment. The total deficiency payment owed is the sum of the individual deficiency owed amounts for each fee amount previously and erroneously paid as a micro entity.

- (3) If the requirements of paragraphs (k)(1) and (2) of this section are not complied with, such failure will either be treated at the option of the Office as an authorization for the Office to process the deficiency payment and charge the processing fee set forth in § **1.17(i)**, or result in a requirement for compliance within a one-month time period that is not extendable under § **1.136(a)** to avoid the return of the fee deficiency payment.

- (4) Any deficiency payment (based on a previous erroneous payment of a micro entity fee) submitted under this paragraph will be treated as a notification of a loss of entitlement to micro entity status under paragraph (i) of this section.

## Certificate of Micro Entity Status – Page 1

Degree of difficulty of filling highlighted items: EASY

Doc Code: MES GIB
Document Description: Certification of Micro Entity Status (Gross Income Basis)

PTO/SB/15A (03-13)

### CERTIFICATION OF MICRO ENTITY STATUS
(GROSS INCOME BASIS)

| Application Number or Control Number (if applicable): | Patent Number (if applicable): |
|---|---|
| 14028786 | |
| First Named Inventor: | Title of Invention: |
| Roy Lique | Digital Camera Lens Guard and Use Extender |

The applicant hereby certifies the following—

(1) **SMALL ENTITY REQUIREMENT** - The applicant qualifies as a small entity as defined in 37 CFR 1.27.

(2) **APPLICATION FILING LIMIT** - Neither the applicant nor the inventor nor a joint inventor has been named as the inventor or a joint inventor on more than four previously filed U.S. patent applications, excluding provisional applications and international applications under the Patent Cooperation Treaty (PCT) for which the basic national fee under 37 CFR 1.492(a) was not paid, and also excluding patent applications for which the applicant has assigned all ownership rights or is obligated to assign all ownership rights as a result of the applicant's previous employment.

(3) **GROSS INCOME LIMIT ON APPLICANTS AND INVENTORS** - Neither the applicant nor the inventor nor a joint inventor, in the calendar year preceding the calendar year in which the applicable fee is being paid, had a gross income, as defined in section 61(a) of the Internal Revenue Code of 1986 (26 U.S.C. 61(a)), exceeding the "Maximum Qualifying Gross Income" reported on the USPTO website at http://www.uspto.gov/patents/law/micro_entity.jsp which is equal to three times the median household income for that preceding calendar year, as most recently reported by the Bureau of the Census.

(4) **GROSS INCOME LIMIT ON PARTIES WITH AN "OWNERSHIP INTEREST"** - Neither the applicant nor the inventor nor a joint inventor has assigned, granted, or conveyed, nor is under an obligation by contract or law to assign, grant, or convey, a license or other ownership interest in the application concerned to an entity that, in the calendar year preceding the calendar year in which the applicable fee is being paid, had a gross income, as defined in section 61(a) of the Internal Revenue Code of 1986, exceeding the "Maximum Qualifying Gross Income" reported on the USPTO website at http://www.uspto.gov/patents/law/micro_entity.jsp which is equal to three times the median household income for that preceding calendar year, as most recently reported by the Bureau of the Census.

**SIGNATURE by a party set forth in 37 CFR 1.33(b)**

| Signature | /Roy Lique/ | | | |
|---|---|---|---|---|
| Name | Roy Lique | | | |
| Date | 2013-09-17 | Telephone | 909-594-1470 | Registration No. |

☐ There is more than one inventor and I am one of the inventors who are jointly identified as the applicant. Additional certification form(s) signed by the other joint inventor(s) are included with this form.

## Certificate of Micro Entity Status – Page 2

Privacy
Statement

### Privacy Act Statement

The Privacy Act of 1974 (P.L. 93-579) requires that you be given certain information in connection with your submission of the attached form related to a patent application or patent. Accordingly, pursuant to the requirements of the Act, please be advised that: (1) the general authority for the collection of this information is 35 U.S.C. 2(b)(2); (2) furnishing of the information solicited is voluntary; and (3) the principal purpose for which the information is used by the U.S. Patent and Trademark Office is to process and/or examine your submission related to a patent application or patent. If you do not furnish the requested information, the U.S. Patent and Trademark Office may not be able to process and/or examine your submission, which may result in termination of proceedings or abandonment of the application or expiration of the patent.

The information provided by you in this form will be subject to the following routine uses:

1. The information on this form will be treated confidentially to the extent allowed under the Freedom of Information Act (5 U.S.C. 552) and the Privacy Act (5 U.S.C 552a). Records from this system of records may be disclosed to the Department of Justice to determine whether disclosure of these records is required by the Freedom of Information Act.
2. A record from this system of records may be disclosed, as a routine use, in the course of presenting evidence to a court, magistrate, or administrative tribunal, including disclosures to opposing counsel in the course of settlement negotiations.
3. A record in this system of records may be disclosed, as a routine use, to a Member of Congress submitting a request involving an individual, to whom the record pertains, when the individual has requested assistance from the Member with respect to the subject matter of the record.
4. A record in this system of records may be disclosed, as a routine use, to a contractor of the Agency having need for the information in order to perform a contract. Recipients of information shall be required to comply with the requirements of the Privacy Act of 1974, as amended, pursuant to 5 U.S.C. 552a(m).
5. A record related to an International Application filed under the Patent Cooperation Treaty in this system of records may be disclosed, as a routine use, to the International Bureau of the World Intellectual Property Organization, pursuant to the Patent Cooperation Treaty.
6. A record in this system of records may be disclosed, as a routine use, to another federal agency for purposes of National Security review (35 U.S.C. 181) and for review pursuant to the Atomic Energy Act (42 U.S.C. 218(c)).
7. A record from this system of records may be disclosed, as a routine use, to the Administrator, General Services, or his/her designee, during an inspection of records conducted by GSA as part of that agency's responsibility to recommend improvements in records management practices and programs, under authority of 44 U.S.C. 2904 and 2906. Such disclosure shall be made in accordance with the GSA regulations governing inspection of records for this purpose, and any other relevant (i.e., GSA or Commerce) directive. Such disclosure shall not be used to make determinations about individuals.
8. A record from this system of records may be disclosed, as a routine use, to the public after either publication of the application pursuant to 35 U.S.C. 122(b) or issuance of a patent pursuant to 35 U.S.C. 151. Further, a record may be disclosed, subject to the limitations of 37 CFR 1.14, as a routine use, to the public if the record was filed in an application which became abandoned or in which the proceedings were terminated and which application is referenced by either a published application, an application open to public inspection or an issued patent.
9. A record from this system of records may be disclosed, as a routine use, to a Federal, State, or local law enforcement agency, if the USPTO becomes aware of a violation or potential violation of law or regulation.

## Electronic Acknowledgement

## Electronic Acknowledgement Receipt – Page 1

| Electronic Acknowledgement Receipt | |
|---|---|
| EFS ID: | 16874420 |
| Application Number: | 14028852 |
| International Application Number: | |
| Confirmation Number: | 8396 |
| Title of Invention: | Digital Camera Lens Guard and Use Extender |
| First Named Inventor/Applicant Name: | Roy Lique |
| Customer Number: | 117033 |
| Filer: | Roy Lique |
| Filer Authorized By: | |
| Attorney Docket Number: | Camera Coup |
| Receipt Date: | 17-SEP-2013 |
| Filing Date: | |
| Time Stamp: | 14:00:31 |
| Application Type: | Utility under 35 USC 111(a) |

**Payment information:**

| Submitted with Payment | no |
|---|---|

**File Listing:**

| Document Number | Document Description | File Name | File Size(Bytes)/ Message Digest | Multi Part /.zip | Pages (if appl.) |
|---|---|---|---|---|---|
| 1 | Certification of Micro Entity (Gross Income Basis) | CetifMicroEntityOfficeOrg.pdf | 130738 | no | 2 |

**Warnings:**

**Information:**

**Electronic Acknowledgement Receipt – Page 2**

| | Total Files Size (in bytes): | 130738 |
|---|---|---|

This Acknowledgement Receipt evidences receipt on the noted date by the USPTO of the indicated documents, characterized by the applicant, and including page counts, where applicable. It serves as evidence of receipt similar to a Post Card, as described in MPEP 503.

**New Applications Under 35 U.S.C. 111**
If a new application is being filed and the application includes the necessary components for a filing date (see 37 CFR 1.53(b)-(d) and MPEP 506), a Filing Receipt (37 CFR 1.54) will be issued in due course and the date shown on this Acknowledgement Receipt will establish the filing date of the application.

**National Stage of an International Application under 35 U.S.C. 371**
If a timely submission to enter the national stage of an international application is compliant with the conditions of 35 U.S.C. 371 and other applicable requirements a Form PCT/DO/EO/903 indicating acceptance of the application as a national stage submission under 35 U.S.C. 371 will be issued in addition to the Filing Receipt, in due course.

**New International Application Filed with the USPTO as a Receiving Office**
If a new international application is being filed and the international application includes the necessary components for an international filing date (see PCT Article 11 and MPEP 1810), a Notification of the International Application Number and of the International Filing Date (Form PCT/RO/105) will be issued in due course, subject to prescriptions concerning national security, and the date shown on this Acknowledgement Receipt will establish the international filing date of the application.

## Notice of Publication

Publication of patent applications is required by the American Inventors Protection Act of 1999 for most plant and utility patent applications filed on or after November 29, 2000. On filing of a plant or utility application on or after November 29, 2000, an applicant may request that the application not be published, but only if the invention has not been and will not be the subject of an application filed in a foreign country that requires publication 18 months after filing (or earlier claimed priority date) or under the Patent Cooperation Treaty. Publication occurs after the expiration of an 18-month period following the earliest effective filing date or priority date claimed by an application. Following publication, the application for patent is no longer held in confidence by the Office and any member of the public may request access to the entire file history of the application.

As a result of publication, an applicant may assert provisional rights. These rights provide a patentee with the opportunity to obtain a reasonable royalty from a third party that infringes a published application claim provided actual notice is given to the third party by applicant, and patent issues from the application with a substantially identical claim. Thus, damages for pre-patent grant infringement by another are now available.

## Notice of Publication of application

The identified application will be electronically published as a patent application publication pursuant to 37 CFR 1.11, et seq. The patent application number and publication date are set forth across.

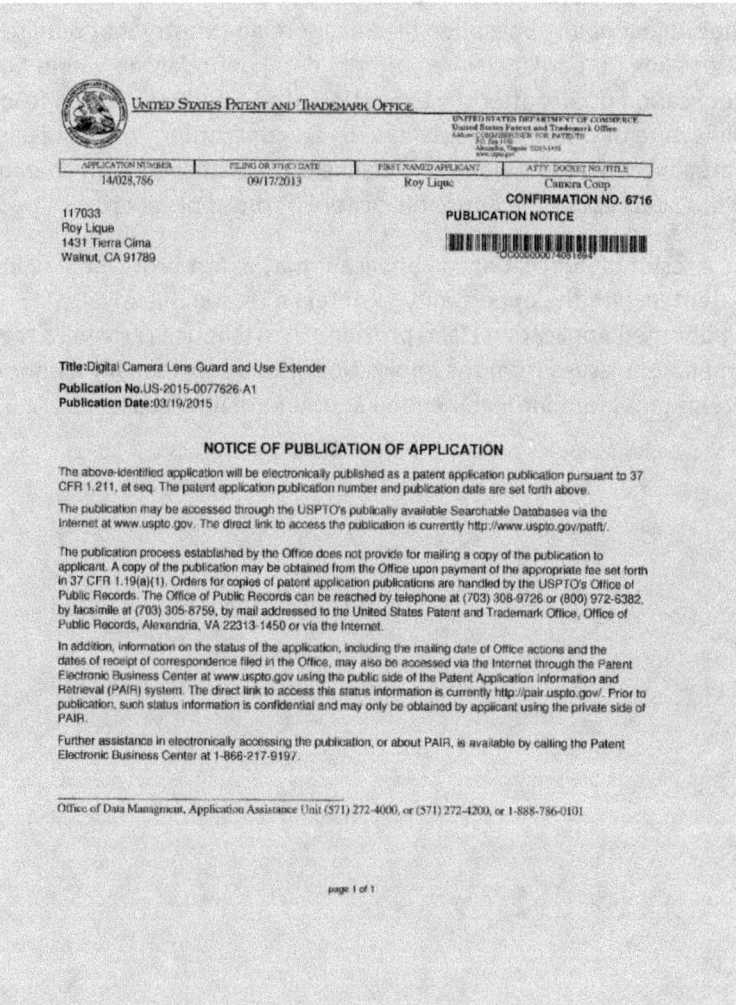

## Publication

Front drawing, abstract

The publication is only a requirement of the law and is not a prelude to final grant of a patent.

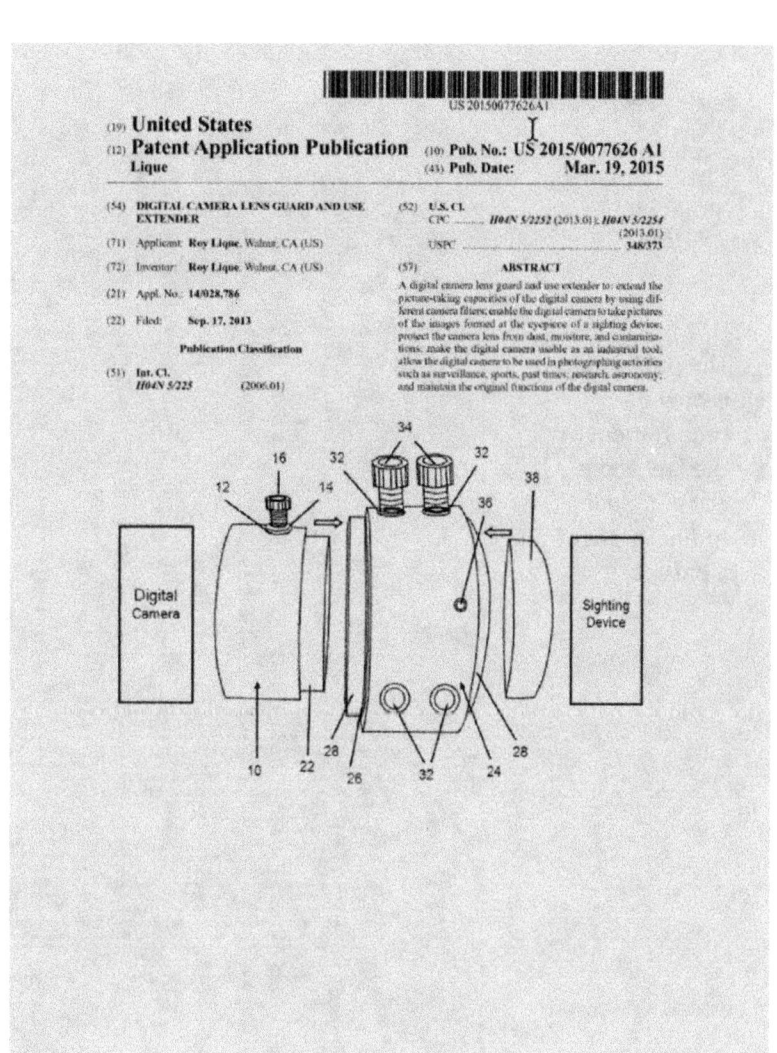

Cross reference, summary

The publication is only a requirement of the law and is not a prelude to final grant of a patent.

US 2015/0077626 A1    Mar. 19, 2015

## DIGITAL CAMERA LENS GUARD AND USE EXTENDER

### CROSS REFERENCE TO RELATED APPLICATIONS

[0001] This application claims the benefit of provisional patent application Ser. No. 61/849,349, filed Jan. 25, 2013 by the present inventor.

### BACKGROUND PRIOR ART

[0002] The following is a tabulation of some prior arts that presently appear barely relevant to the embodiments of the digital camera lens guard and use extender.

#### U.S. PATENTS

[0003]

| Pat. No. | Title | Issue Date | Patentee |
|---|---|---|---|
| 8,111,294 | Matte lens assembly | Feb. 7, 2012 | Wood; Dennis |
| 8,054,545 | Lens hood for a camera lens | Nov. 8, 2011 | Cheng; Klaxy Cheng |
| 7,813,039 | Camera cover | Oct. 12, 2010 | Yuntay; Ozzine |
| 6,243,540 | Lens barrel assembly | Jun. 5, 2001 | Kume; Katsuhi |
| 5,136,481 | Lens hood for a photographic lens | Jun. 30, 1992 | Croνα; Jeff |
| 4,533,212 | Accessory holding device for optical instrument | Aug. 6, 1985 | Shimizu; Sachio |
| 4,384,767 | Clamping device for camera accessory or hood | May 24, 1983 | Kawai; Isheo |
| 4,295,776 | Combined lens cap and sunshade for a camera | Oct. 20, 1981 | Frost; George H. |
| 4,137,560 | Camera matte box | Jan. 30, 1979 | Chrfar; Jack |
| 3,906,107 | Hood for the lens of optical instruments with pivotally mounted lens cover | Sep. 30, 1975 | Neuthert; Jody L. |
| 3,840,883 | CAMERA LENS HOOD | Oct. 8, 1974 | Chuen; J. Robert |
| 3,614,196 | COMBINED LENS HOOD AND FILTER SUPPORT | Oct. 19, 1971 | Schlager; Werner |

[0004] In the absence of significant relevance between prior arts and the embodiments of the digital camera lens guard and use extender, the use and benefits of the latter are discussed to create and present a new line of products principally dedicated to digital cameras.

[0005] The embodiments of the digital camera lens guard and use extender are in the field of cameras. More particularly, the embodiments extend and expand the capacities of a digital camera while also protecting the camera's lens. Capacities expansion is done with older as well as newer hardware, or a combination of both. Where applicable, expansion will also be done with software.

[0006] While the camera market is flooded with new camera models every year, it is also being depleted by obsolescence due to incompatibility with new software and hardware. It is also depleted by more dislikes of older models and the accessories that come with them. Phone cameras also diminish the popularity of digital cameras. Some victims of obsolescence include multi-image lenses, fisheye lenses, macro and telephoto lenses, square-shaped filters, over-sized and under-sized filters, fog and snow lenses, so on so forth.

[0007] Interchangeability of accessories between the digital camera and certain models of expensive cameras, is rare. The digital camera's lack of threads accounts for threaded accessories being almost exclusive monopolies of certain camera models. Some exciting photographs are taken using threaded accessories. Making the interchangeability problem more obvious is the fact that threads come in either metric or English.

[0008] Notwithstanding the additional sophisticated features coming with new cameras, there are still missed capacities that one would like to have in the digital camera. Examples are documenting an event happening too far from the viewer, or a past time too dangerous for the viewer to come close to the subject.

[0009] The additional new features that come with the later digital camera models include expensive electronics that need more protection. The entry of dust, moisture, and other contaminations into the lens area must be minimized in order to maintain the digital camera in an efficient working order. By engaging an embodiment of the digital camera lens guard and use extender to the digital camera and screwing in a camera filter, ample protection of the camera electronics is already provided.

[0010] The digital camera equipped with the embodiment of the digital camera lens guard and use extender, meets both the camera users' needs for additional camera capacities and lens and electronics protection without investing in expensive equipments.

### SUMMARY

[0011] Functionally, the digital camera equipped with an embodiment of the digital camera lens guard and use extender compares with the more expensive models. The embodiment concentrates on enhancing the capacities of the digital camera and protecting its electronics.

[0012] With different size adapter rings at the ends of the primary ring of the embodiment, a significant number of camera filters can be used. Obsolete and old camera filters are given new life because they can be used again. With proper adaptations, lenses from other camera models may now be used with the digital camera.

[0013] In conjunction with telescopes, binoculars, microscopes and other sighting devices, the digital camera can serve as a tool in industrial applications. More photo oppor-

Drawings, first embodiment detailed description

The publication is only a requirement of the law and is not a prelude to final grant of a patent.

First embodiment detailed description, continued

The publication is only a requirement of the law and is not a prelude to final grant of a patent.

US 2015/0077626 A1

Mar. 19, 2015

3

[0039]  Slip Type Camera Mount Assembly.

[0040]  In more details, still referring to FIG. 1 and FIG. 2, a slip type camera mount assembly comprises the anchor ring 10, an anchor hole 12, an anchor nut 14, an anchor screw 16, and a barrel track 22.

[0041]  The action of slipping the barrel track 22 into the anchor ring 10 after the anchor ring 10 is mounted on the digital camera, is what "slip type" camera mount assembly refers to. The action, simultaneous with slipping the other end of the barrel 22 to the primary ring 24, secures the embodiment of the digital camera lens guard and use extender to the digital camera.

[0042]  Anchor Ring 10.

[0043]  The cylindrical anchor ring 10 of approximately 13 mm (0.512") in length, with approximately 44.7 mm (1.759") inside diameter, and with approximately 1.5 mm (0.058") wall is cut from 6061 grade aluminum. The wall provides sufficient thickness for attaching the anchor ring 10 upright to the digital camera concentrically with the camera barrel. The anchor hole 12 facilitates the attachment of a pair of sufficiently large anchor screw 16 and anchor nut 14, to the anchor ring 10. The anchor screw 16 and the anchor nut 14 are used to secure the barrel track 22.

[0044]  Barrel Track 22.

[0045]  The cylindrical barrel track 22 is also from 6061 grade aluminum, cut to an approximate length of 25.4 mm (1.0"). Approximately one half of its entire length is inserted into the anchor ring 10 to support a secure connection. The other approximate half serves as a flange that inserts into the primary ring 24. The depth of insertion of the flange into the primary ring 24 is adjustable, making it useful in minimizing the formation of circles around pictures taken by the digital camera.

[0046]  Main Housing Assembly.

[0047]  In more details still referring to FIG. 1 and FIG. 2, the main housing assembly comprises a cylindrical primary ring 24 with appurtenances. It is in the main housing assembly that the major components of the embodiment of the digital camera lens guard and use extender come together and form a secure connection. Photographing activities take place in the main housing assembly.

[0048]  Primary Ring 24.

[0049]  The primary ring 24 accepts and secures a camera filter or a sighting device through the embedded inner adapter ring 30 (not shown) or outer adapter ring 28 at its ends. Additionally to FIG. 1, the primary ring 24 accepts and secures a sighting device through a coupler ring 38.

[0050]  The primary ring 24 is also cut from 6061 grade aluminum to approximately 25.4 mm (1.0") long. Sufficient space for embedding the commonly used size ranges of adapter rings is provided by its approximate outside diameter of 63.5 mm (2.5") and approximate wall of 6.4 mm (0.252"). Its approximate inside diameter of 44.5 mm (1.752") provides the additional track for the camera barrel to extend and retract without obstruction.

[0051]  Owing to the identical ends of the primary ring 24 with different size adapter rings, the directions at which the ends point are reversible, offering more opportunities for attaching different camera filters or sighting devices, one at a time.

[0052]  Adapter Rings.

[0053]  Size-matched and gender-matched camera filter or sighting device is accepted at either end of the primary ring 24. The embedded adapter rings at the ends of the primary ring 24 immediately provide more opportunities to use different camera filters and sighting devices, one at a time. Since circles around images taken by the digital camera are sometimes caused by the addition of loose adapter rings and camera filters, care is observed that only enough of them are used as needed.

[0054]  The inside and outside diameters of a target adapter ring are modified as necessary to sufficient sizes so that the adapter ring can be embedded at either end of the primary ring 24. Optionally, its male threads are stripped off. Two types of adapter rings are optionally embedded at either end of the primary ring 24, namely, inner adapter ring 30 and outer adapter ring 28. Their construction details are described as follows:

[0055]  Inner Adapter Ring 30.

[0056]  To attach the inner adapter ring 30 at either end of the primary ring 24, an approximately 6.4 mm (0.252") deep recess 26 with sufficient circumference to accept the target inner adapter ring 30, is curved. The inner adapter ring 30 is attached resting at the bottom of the recess 26 with the female threads oriented outwards, using industry grade adhesive.

[0057]  Outer Adapter Ring 28.

[0058]  To attach the outer adapter ring 28 at either end of the primary ring 24, an optional approximately 3.2 mm (0.125") deep recess 26 with sufficient circumference to accept the target outer adapter ring 28, is curved. The outer adapter ring 28 is attached resting at the bottom of the recess 26 with the female threads oriented outwards, using industry grade adhesive.

[0059]  An alternative way to embed the outer adapter ring 28 into the primary ring 24 is to closely cut off the recessed portion of the primary ring 24, referring to previous paragraph. Using industry grade adhesive, the outer adapter ring 28 is inserted and attached flushed with the end of the cut off portion. The cut off portion is attached back to the primary ring 24 making sure the female threads are oriented outwards.

[0060]  The outer adapter ring 28 can also be attached directly to either end of the primary ring 24 without curving the recess 26. It only needs to be stripped off of its male threads and attached to the primary ring 24 with the female threads oriented outwards, using industry grade adhesive.

[0061]  Stack of Adapter Rings.

[0062]  Building a stack of adapter rings as shown in FIG. 7, follows the steps described for embedding the inner adapter ring 30 and the outer adapter ring 28. The inner adapter rings 30 progressively get smaller as the stack grows, each inner adapter ring 30 resting approximately 6.4 mm (0.252") deeper from the preceding larger one.

[0063]  Retaining Holes 32 and Retaining Screws 34.

[0064]  In order to provide ample handling surface, retaining holes 32 of sufficient size are drilled approximately 6.4 mm (0.252") from either end of the primary ring 24, because at least one retaining screw 34 is used to secure the barrel track 22 and another one to secure the coupler ring 38, the locations of the retaining holes 32 are sufficiently far apart to allow the retaining screws 34 to turn freely.

[0065]  Insert Holes 36.

[0066]  The optional insert holes 36 are of sufficient size to accept inserts for expansion and improvements. They are drilled onto the primary ring 24 at desired locations that do not interfere with the outer adapter rings 28, inner adapter rings 30, and retaining screws 34.

# First embodiment operation, additional embodiments

The publication is only a requirement of the law and is not a prelude to final grant of a patent.

US 2015/0077626 A1

Mar. 19, 2015

4

[0067] Coupler Ring 38.

[0068] In more details referring to FIG. 1, the coupler ring 38 provides the connection between the main housing assembly and the sighting device. The sighting device is secured to the main housing assembly by properly inserting its eyepiece into the coupler ring 38 and the coupler ring 38 into the primary ring 24.

[0069] The coupler ring 38 is of sufficient size and length, preferably between 25.4 mm (1.0") and 50.8 mm (2.0"), and has an outside diameter closely matching the inside diameter of the primary ring 24. The inside diameter of the coupler ring 38 varies depending on the size of the target eyepiece of the sighting device. If necessary, the coupler ring 38 is machined to alter its inside diameter in order to match it with the size of the eyepiece of the sighting device. The coupler ring 38 is also cut from 6061 grade aluminum.

[0070] The construction details of the embodiments of the digital camera lens guard and use extender as shown in FIG. 1 and FIG. 2 are that the embodiments may be made of metal or of any other sufficiently rigid and strong material such as high-strength plastic and the like. Further, the various components of the embodiments of the digital camera lens guard and use extender can be made of different materials from different sources, brands, and styles.

Operation

FIG. 1 and FIG. 2

[0071] Referring to FIG. 1 and FIG. 2, either end of the barrel track 22 is inserted into the anchor ring 10. Anchor screw 16 is tightened to secure the barrel track 22.

[0072] The open end of the barrel track 22 is inserted into either end of the primary ring 24. It is inserted opposite the end where the camera filter or sighting device is attached. The size and type of camera filter or sighting device determine which end of the primary ring 24 needs to accept and secure the barrel track 22.

[0073] Insertion depth of the barrel track 22 into the primary ring 24 is fixed for the distance the camera barrel has to extend. One or more retaining screws 34 are tightened to secure the barrel track 22.

[0074] In more details referring to FIG. 1, the eyepiece of a sighting device is inserted into either end of the coupler ring 38. If the outside circumference of the eyepiece is smaller than the inside circumference of the coupler ring 38, a fitting (not shown) is used to make the insertion snugly secure. The open end of the coupler ring 38 is inserted as far as it can go or until it is stopped by the barrel track 22, into the primary ring 24.

[0075] Alternatively, the use of the coupler ring 38 can be done away with as shown in FIG. 2. Size-matched and gender-matched camera filter or sighting device is screwed directly to either the outer adapter ring 28 or the inner adapter ring 30 (not shown). If necessary, loose adapter rings are added to find a match between the embedded adapter rings and the target camera filter or sighting device.

[0076] In more details referring to FIG. 1 and FIG. 2, as shown, the embodiments of the digital camera lens guard and use extender provide for quick and easy installation of the camera filter or the sighting device to the main housing assembly. They also facilitate their quick and easy mounting and dismounting to and from the digital camera. More size-matched and gender-matched camera filters and sighting

devices unusable with the digital camera before, become available now with the use of the outer adapter rings 28 and the inner adapter rings 30.

FIG. 3 and FIG. 4

Additional Embodiments

[0077] Referring now to FIG. 3 and FIG. 4, there are shown embodiments of the digital camera lens guard and use extender having a camera mount assembly, a main housing assembly, and additionally in the case of FIG. 3, a coupler ring 38. The figures further show the position of each embodiment in relation to the digital camera represented by boxed "Digital Camera", and to the attachments represented by boxed "Sighting Device" or boxed "Camera Filter/Sighting Device". When used with the coupler ring 38, the embodiment offers only a single attachment and its position is shown in FIG. 3. With multiple allowable attachments attached one at a time, the position of the embodiment is shown in FIG. 4. The only difference between the embodiments shown in FIG. 3 and FIG. 4 is the presence or absence of the coupler ring 38.

[0078] Descriptions of the functions and construction details of the shown embodiments of the digital camera lens guard and use extender follow.

[0079] Camera Mount Assembly.

[0080] Still referring to FIG. 3 and FIG. 4, a base adapter ring 18 of the camera mount assembly secures the embodiment of the digital camera lens guard and use extender to the digital camera. It is the basis for most of the sizes, values, measurements, and dimension of the other components of the embodiments.

[0081] Screw Type Camera Mount Assembly.

[0082] In more details still referring to FIG. 3 and FIG. 4, the screw type camera mount assembly comprises the modified base adapter ring 18 and the configured barrel track 22.

[0083] The action of screwing the track mounting ring 20 to the modified base adapter ring 18 after the base adapter ring 18 is mounted on the digital camera, is what "screw type" camera mount assembly refers to. The action, simultaneous with slipping the other end of the barrel 22 to the primary ring 24, secures the embodiment of the digital camera lens guard and use extender to the digital camera.

[0084] Base Adapter Ring 18.

[0085] Initially, the inside diameter of the base adapter ring 18 is approximately 40 mm (1.575") to 45 mm (1.772") and its outside diameter is approximately 50 mm (1.969"). The sizes are suitable for modification by machining, to circumferentially enclose the camera barrel. With its male threads stripped off, the base adapter ring 18 is attached to the digital camera upright concentrically with the camera barrel and with the female threads oriented outwards, using industry grade adhesive. To allow the camera barrel to extend and retract without obstruction, a space is maintained between the camera barrel and the base adapter ring 18.

[0086] Barrel Track 22.

[0087] The barrel track 22 is also cut from 6061 grade aluminum to an approximate length of 25.4 mm (1.0"). Using industry grade adhesive, one end is fitted with the modified track mounting ring 20 which is size-matched and gender-matched with the base adapter ring 18. The track mounting ring 20 is made flush with the end of the barrel track 22. The other end serves as a flange that inserts into the primary ring 24. The depth of insertion of the flange into the primary ring

Additional embodiments operation, alternative embodiments

The publication is only a requirement of the law and is not a prelude to final grant of a patent.

US 2015/0077626 A1

Mar. 19, 2015

5

24 is adjustable, making it useful in minimizing the formation of circles around pictures taken by the digital camera.

[0088] Main Housing Assembly.

[0089] The main housing assembly is identical to that described in the embodiments shown in FIG. 1 and FIG. 2.

[0090] Coupler Ring 38.

[0091] The coupler ring 38 is identical to that described in the embodiments shown in FIG. 1 and FIG. 2.

[0092] The construction details of the embodiments of the digital camera lens guard and use extender as shown in FIG. 3 and FIG. 4 are that the embodiments may be made of metal or of any other sufficiently rigid and strong material such as high-strength plastic and the like. Further, the various components of the embodiments of the digital camera lens guard and use extender can be made of different materials from different sources, brands, and styles.

### Operation

#### FIG. 3 and FIG. 4

[0093] Referring to FIG. 3 and FIG. 4, the track mounting ring 20 is screwed into the base adapter ring 18.

[0094] The open end of the barrel track 22 is inserted into either end of the primary ring 24. It is inserted opposite the end where the camera filter or sighting device is attached. The size and type of camera filter or sighting device determine which end of the primary ring 24 needs to accept and secure the barrel track 22.

[0095] Insertion depth of the barrel track 22 into the primary ring 24 is fixed for the distance the camera barrel has to extend. One or more retaining screws 34 are tightened to secure the barrel track 22.

[0096] In more details referring to FIG. 3, the eyepiece of a sighting device is inserted into either end of the coupler ring 38. If the outside circumference of the eyepiece is smaller than the inside circumference of the coupler ring 38, a fitting (not shown) is used to make the insertion snugly secure. The open end of the coupler ring 38 is inserted as far as it can go or until it is stopped by the barrel track 22, into the primary ring 24.

[0097] Alternatively, the use of the coupler ring 38 can be done away with as shown in FIG. 4. Size-matched and gender-matched camera filter or sighting device is screwed directly to either the outer adapter ring 28 or the inner adapter ring 30 (not shown). If necessary, lower adapter rings are added to find a match between the embedded adapter rings and the target camera filter or sighting device.

[0098] In more details referring to FIG. 3 and FIG. 4, as shown, the embodiments of the digital camera lens guard and use extender provide for quick and easy installation of the camera filter or the sighting device to the main housing assembly. They also facilitate their quick and easy mounting and dismounting to and from the digital camera. More size-matched and gender-matched camera filters and sighting devices unusable with the digital camera before become available now with the use of the outer adapter rings 28 and the inner adapter rings 30.

#### FIG. 5 and FIG. 6

##### Alternative Embodiments

[0099] Referring to FIG. 5 and FIG. 6, there are shown simplified embodiments of the digital camera lens guard and use extender. The figures further show the position of each

embodiment in relation to the digital camera represented by boxed "Digital Camera". With multiple allowable attachments represented by boxed "Camera Filter/Sighting Device" attached one at a time, the relative positions of the embodiments are also shown in both figures.

[0100] In FIG. 5 the end of the barrel track 22 opposite the end that slips into the anchor ring 10 is fitted with a track end ring 40 with the female threads oriented outwards. In FIG. 6, fitting is done at the end opposite that which screws into the base adapter ring 18. In both instances, fitting is done using industry grade adhesive and ensuring that the track end ring 40 is flushed with the end of the barrel track 22. The camera filter or sighting device is screwed directly into the track end ring 40.

[0101] The construction details of the embodiments of the digital camera lens guard and use extender as shown in FIG. 5 and FIG. 6 are that the embodiments may be made of metal or of any other sufficiently rigid and strong material such as high-strength plastic and the like. Further, the various components of the embodiments of the digital camera lens guard and use extender can be made of different materials from different sources, brands, and styles.

[0102] Referring to FIG. 5 and FIG. 6, as shown, the embodiments of the digital camera lens guard and use extender provide for simple and direct use of camera filters and sighting devices by completely bypassing the main housing assembly.

### Advantages

[0103] Broadly, from the description above, a number of advantages of most embodiments of the digital camera lens guard and use extender become evident:

[0104] (a) The embodiments of the digital camera lens guard and use extender have the advantage of possibly being one of the few dedicated to digital cameras.

[0105] (b) Due to its simple design, future changes on the embodiments of the digital camera lens guard and use extender will be easily implemented.

[0106] (c) The features added by sighting devices make the digital camera more adaptable to various photo opportunities.

[0107] (d) With the added features of a sighting device, the digital camera can be used as an industrial tool.

[0108] (e) Mounting and dismounting of an embodiment to and from the digital camera takes only few turns of the track mounting ring or the anchor screw.

[0109] (f) The digital camera remains portable despite the addition of a base ring.

[0110] (g) Embodiments of the digital camera lens guard and use extender of different sizes can be manufactured for different digital camera types, models, and sighting devices, provided appropriate matching adapter rings are used.

[0111] (h) The digital cameras are now able to take pictures previously possible only with the more expensive cameras.

[0112] (i) Camera protection extends the life of the digital camera and is accomplished with just a few turns of a camera filter.

[0113] (j) Camera users will benefit from innovations and improvements from two different industrial classifications, namely, digital cameras and sighting devices.

Conclusion, ramifications, and scope, claims

The publication is only a requirement of the law and is not a prelude to final grant of a patent.

US 2015/0077626 A1

Mar. 19, 2015

6

[0114] (k) With the different size adapter rings fitted at the ends of the primary ring, immediately a large number of camera filters becomes available.

[0115] (l) Tripods are used less frequently because of the nature of digital cameras.

### CONCLUSION, RAMIFICATIONS, AND SCOPE

[0116] Accordingly, a digital camera user will see that a digital camera enabled by the embodiments of the digital camera lens guard and use extender is well adapted to various photo taking sessions. As the embodiments are easy to mount and dismount to and from the digital camera, more photographic events can be recorded.

[0117] In terms of functionalities, the wider availability of camera filters and sighting devices makes the digital camera comparable with the more expensive types and models. With the option to choose which end of the main housing assembly to attach camera filters and sighting devices, the possibilities become more numerous.

[0118] More specifically, the following are few examples of the use of the digital camera enabled by the embodiments of the digital camera lens guard and use extender:

[0119] surveillance from a distance when getting close endangers the observer.

[0120] observing a phenomenon such as unusual celestial events.

[0121] research such as observing the habits of certain insects.

[0122] crowd observation such as in parades and demonstrations.

[0123] monitoring such as vehicles behaving erratically.

[0124] past time such as bird-watching and whale-watching.

[0125] safaris such as observing wild animals from a distance.

[0126] emergency reporting such as a traffic accident.

[0127] fire fighting such as reporting a fire at a distant canyon.

[0128] progress report such as of a mountain climber.

[0129] fast sports such as tennis matches and basketball games.

[0130] traffic surveillance such as monitoring speeding cars.

[0131] weather observation such as monitoring snow levels.

[0132] crowd safety such as life guarding on the beach.

[0133] law enforcement such as unruly crowd.

[0134] While the foregoing written descriptions of the embodiments of the digital camera lens guard and use extender enables one of ordinary skill to make and use what is considered presently to be the best mode thereof, those of ordinary skills will understand and appreciate the existence of variations, combinations, and equivalents of the specific embodiment, method, and examples herein. The digital cam-

era lens guard and use extender should therefore not be limited by the above described embodiments, methods, and examples, but by all embodiments and methods within the scope and spirit of the digital camera lens guard and use extender as claimed.

I claim:

1. a digital camera lens guard and use extender for adding deviant picture taking capacities to a digital camera, and protecting said camera's lens and electronics, comprising,

a. a camera mount assembly having a circular base and a cylindrical ring of sufficient size and length, for providing enclosure and track for said camera's barrel.

b. a main housing assembly having a cylindrical ring of sufficient size and length with a plurality of retaining screws and adapter rings, for accepting and securing matching camera filters and sighting devices,

2. the digital camera lens guard and use extender of claim 1, wherein said camera mount assembly having a locking method, for engaging said circular base and said cylindrical ring together.

3. the digital camera lens guard and use extender of claim 2 wherein said locking method is of slip type.

4. the digital camera lens guard and use extender of claim 2 wherein said locking method is of screw type.

5. the digital camera lens guard and use extender of claim 1 wherein said cylindrical ring of said camera mount assembly being fitted with an adapter ring, for securing camera filters or sighting devices.

6. the digital camera lens guard and use extender of claim 1 wherein said cylindrical ring of said main housing assembly having more than one embed locations, for securing said adapter rings of said main housing assembly.

7. the digital camera lens guard and use extender of claim 6 wherein said embed location is below the surface at the end of said cylindrical ring.

8. the digital camera lens guard and use extender of claim 6 wherein said embed location is on the surface at the end of said cylindrical ring.

9. the digital camera lens guard and use extender of claim 6 wherein said embed location is in-between a cut off portion of said cylindrical ring and said cylindrical ring itself.

10. the digital camera lens guard and use extender of claim 1 wherein said cylindrical ring of said main housing assembly being inserted with a coupler ring, for securing the eyepiece of a sighting device.

whereby the digital camera lens guard and use extender adds picture taking capacities that are not normally common with a digital camera; provides protection for the camera's lens and electronics; facilitates the use of camera filters, adapter rings, and sighting devices; enables the digital camera to take pictures normally; enables the camera to take pictures of images formed at the eyepiece of a sighting device; and prepares itself for future changes and improvements.

* * * * *

Drawing - Figure 1

The publication
is only a
requirement of
the law and is
not a prelude
to final grant of
a patent.

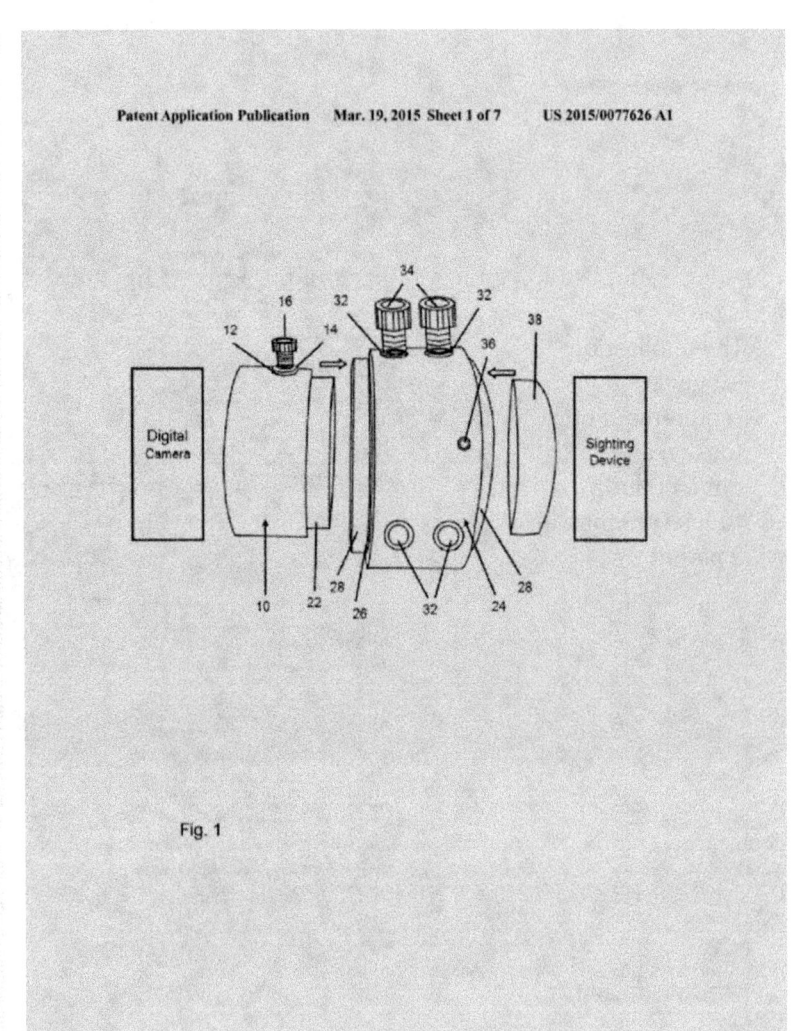

Patent Application Publication    Mar. 19, 2015  Sheet 1 of 7    US 2015/0077626 A1

Fig. 1

Drawing - Figure 2

The publication
is only a
requirement of
the law and is
not a prelude
to final grant of
a patent.

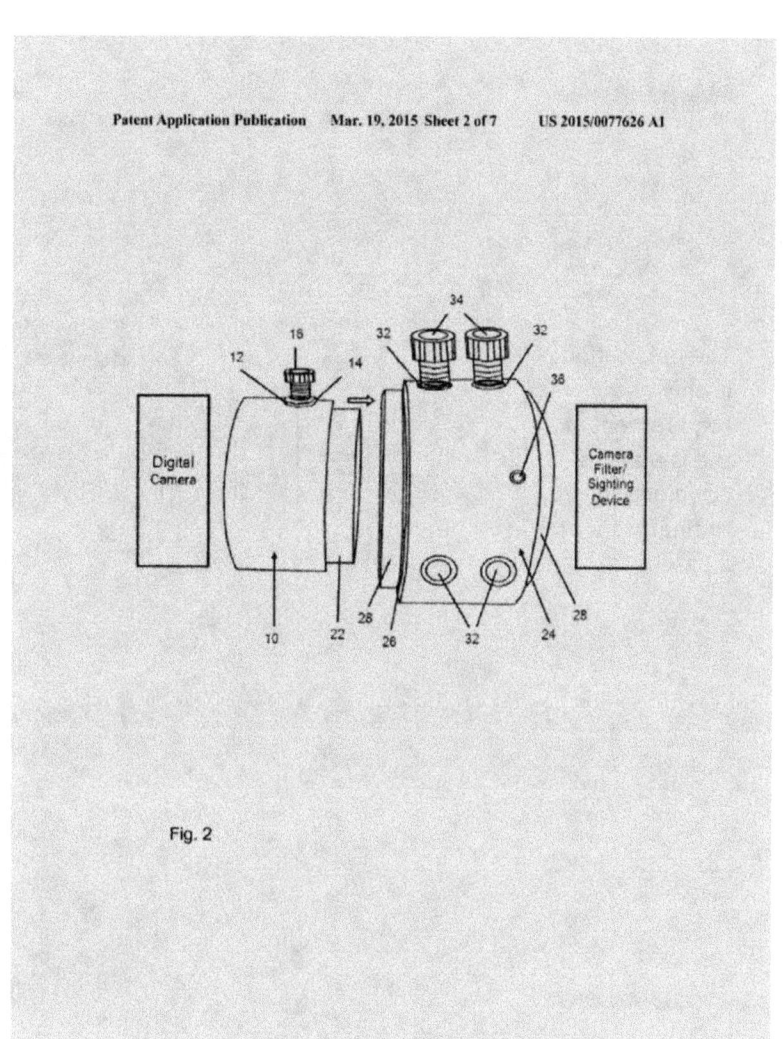

Fig. 2

Drawing - Figure 3

The publication is only a requirement of the law and is not a prelude to final grant of a patent.

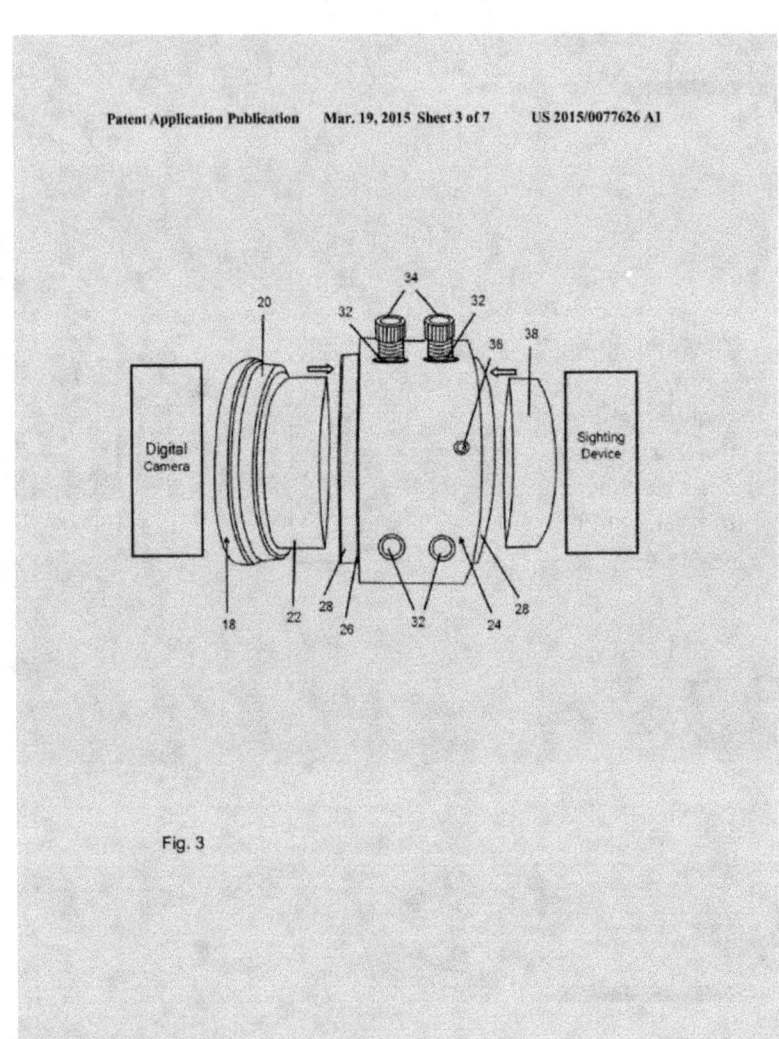

Fig. 3

Drawing - Figure 4

The publication
is only a
requirement of
the law and is
not a prelude
to final grant of
a patent.

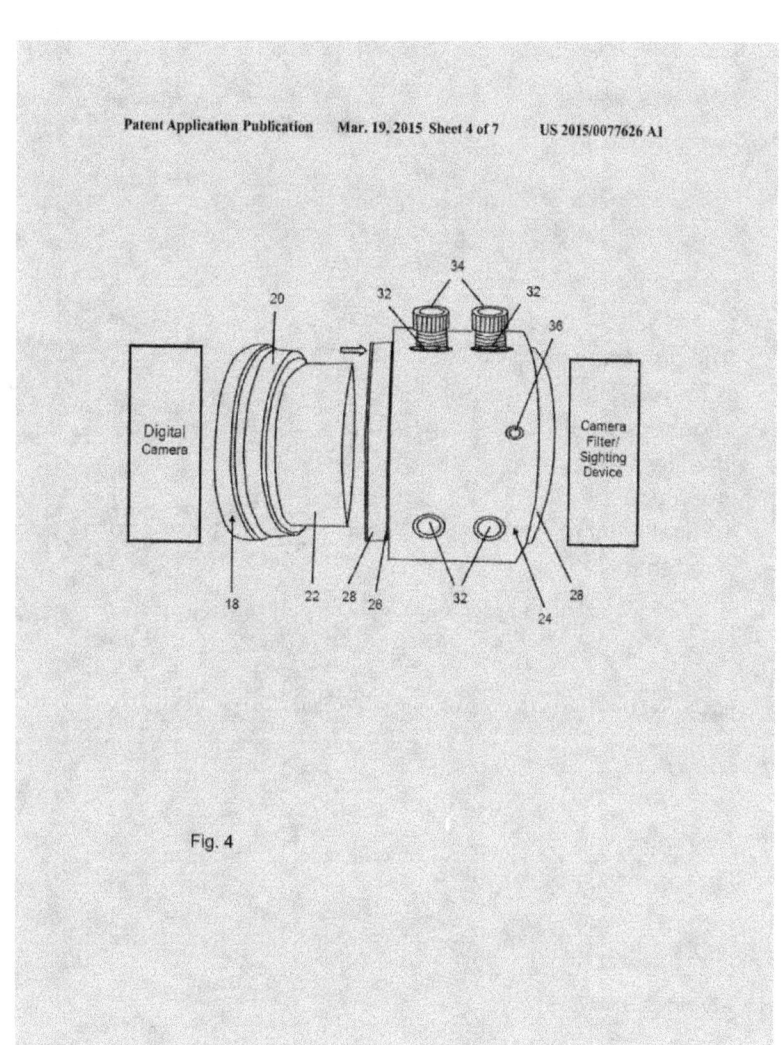

Drawing - Figure 5

The publication
is only a
requirement of
the law and is
not a prelude
to final grant of
a patent.

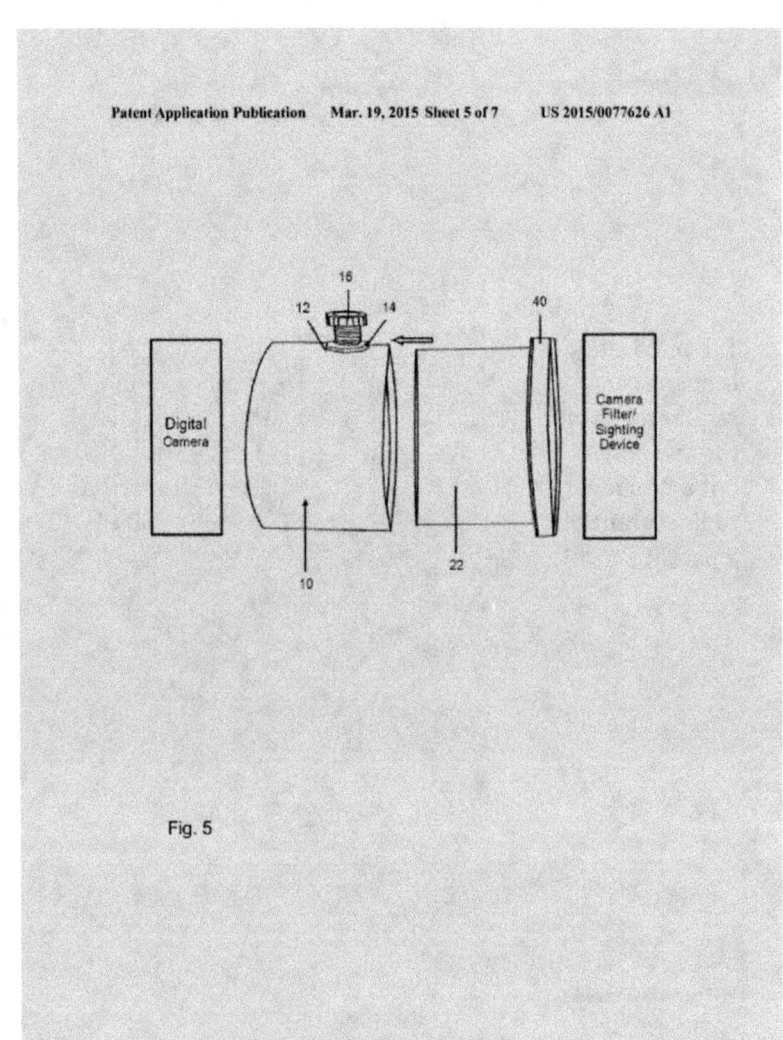

Drawing - Figure 6

The publication
is only a
requirement of
the law and is
not a prelude
to final grant of
a patent.

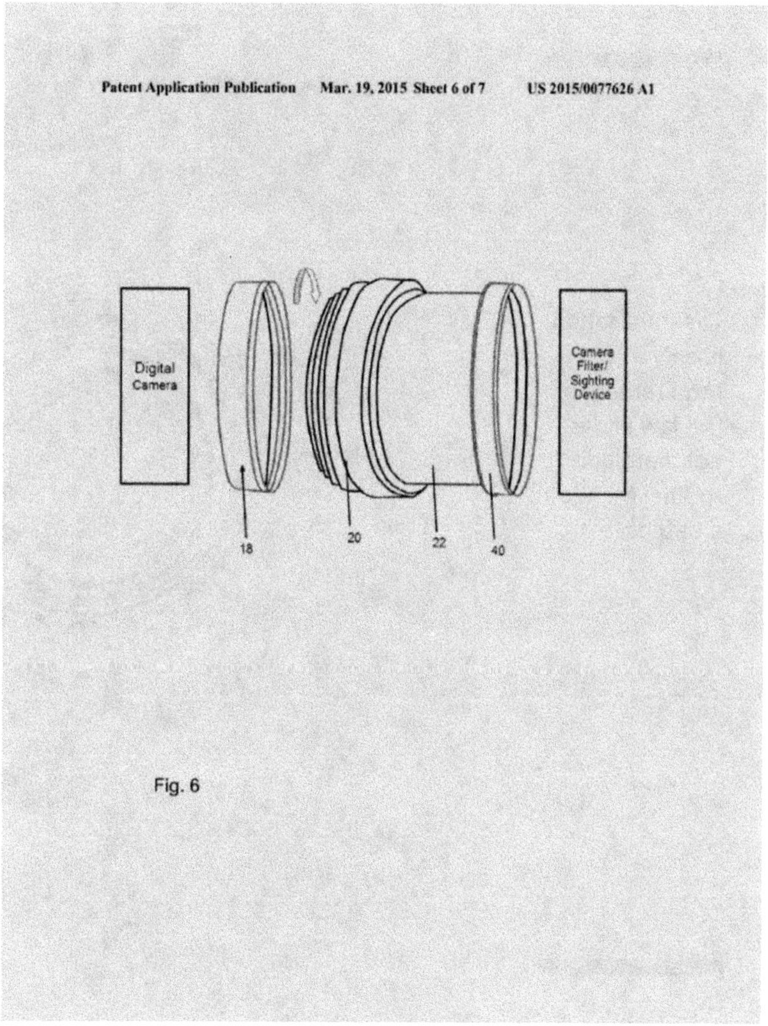

Patent Application Publication     Mar. 19, 2015  Sheet 6 of 7     US 2015/0077626 A1

Digital Camera

Camera Filter/ Sighting Device

18   20   22   40

Fig. 6

Drawing - Figure 7

The publication
is only a
requirement of
the law and is
not a prelude
to final grant of
a patent.

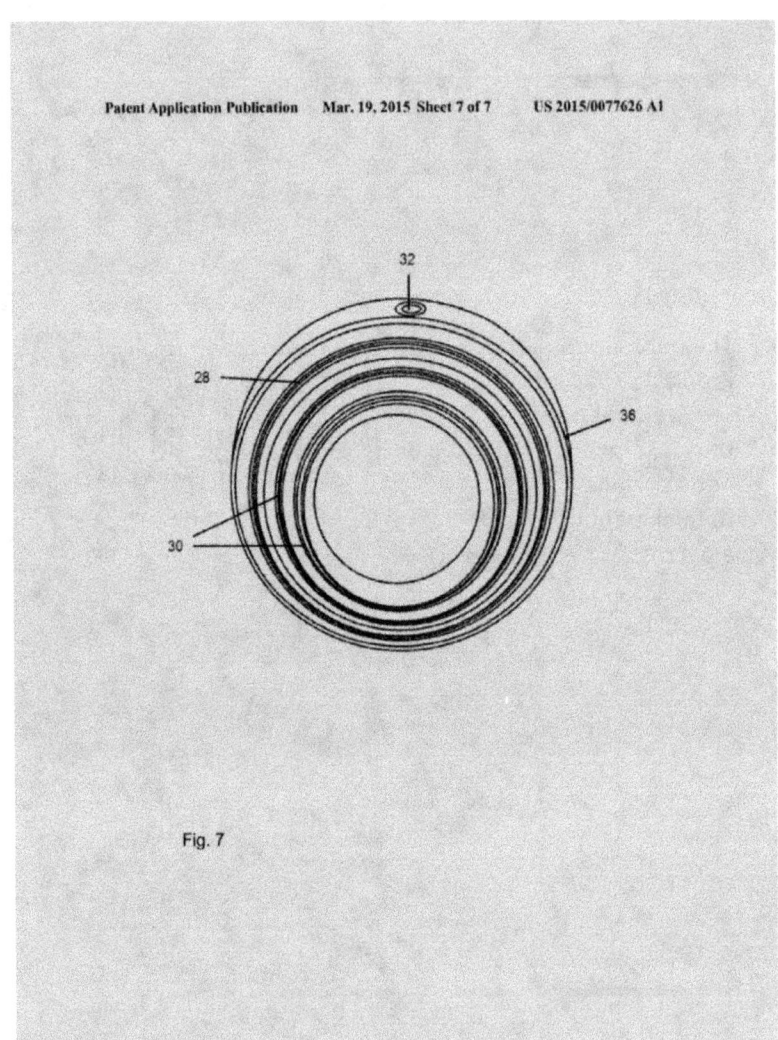

Patent Application Publication    Mar. 19, 2015  Sheet 7 of 7      US 2015/0077626 A1

Fig. 7

206

# Part 6 – Prosecution of Application

## Office Action

The applicant is notified in writing of the examiner's decision by an Office "action" which is normally mailed to the attorney or agent of record. The reasons for any adverse action or any objection or requirement are stated in the Office action and such information or references are given as may be useful in aiding the applicant to judge the propriety of continuing the prosecution of his or her application.

If the claimed invention is not directed to patentable subject matter, the claims will be rejected. If the examiner finds that the claimed invention lacks novelty or differs only in an obvious manner from what is found in the prior art, the claims may also be rejected. It is not uncommon for some or all of the claims to be rejected on the first Office action by the examiner; relatively few applications are allowed as filed.

### Applicant's Reply

The applicant must request reconsideration in writing, and must distinctly and specifically point out the supposed errors in the examiner's Office action. The applicant must reply to every ground of objection and rejection in the prior Office action. The applicant's reply must appear throughout to be a bona fide attempt to advance the case to final action or allowance. The mere allegation that the examiner has erred will not be received as a proper reason for such reconsideration.

In amending an application in reply to a rejection, the applicant must clearly point out why he or she thinks the amended claims are patentable in view of the state of the art disclosed by the prior references cited or the objections made. He or she must also show how the claims as amended avoid such references or objections. After reply by the applicant, the application will be reconsidered, and the applicant will be notified as to the status of the claims—that is, whether the claims are rejected, or objected to, or whether the claims are allowed, in the same manner as after the first examination. The second Office action usually will be made final.

Interviews with examiners may be arranged, but an interview does not remove the necessity of replying to Office actions within the required time.

### Final Rejection

On the second or later consideration, the rejection or other action may be made final. The applicant's reply is then limited to appeal in the case of rejection of any claim and further amendment is restricted. Petition may be taken to the Director in the case of objections or requirements not involved in the rejection of any claim. Reply to a final rejection or action must include cancellation of, or appeal from the rejection of, each claim so rejected and, if any claim stands allowed, compliance with any requirement or objection as to form. In making such final rejection, the examiner repeats or states all grounds of rejection then considered applicable to the claims in the application.

## Amendments to Application

The applicant may amend the application as specified in the rules, or when and as specifically required by the examiner.

Amendments received in the Office on or before the mail date of the first Office action are called "preliminary amendments," and their entry is governed by 37 CFR 1.115. Amendments in reply to a non-final Office action are governed by CFR 1.111. Amendments filed after final action are governed by 37CFR 1.116 and 37CFR 41.33.

The specification, claims, and drawing must be amended and revised when required, to correct inaccuracies of description and definition or unnecessary words, and to provide substantial correspondence between the claims, the description, and the drawing. All amendments of the drawings or specification, and all additions thereto must not include new matter beyond the original disclosure. Matter not found in either, involving a departure from or an addition to the original disclosure cannot be added to the application even if supported by a supplemental oath or declaration, and can be shown or claimed only in a separate application.

The manner of making amendments to an application is provided in 37 CFR 1.121. Amendments to the specification (but not including the claims) must be made by adding, deleting or replacing a paragraph, by replacing a section, or by a substitute specification, as provided in the rules. Replacement paragraphs are to include markings (e.g., underlining and strikethrough) to show all changes relative to the previous version of the paragraph. New paragraphs are to be provided without any underlining. If a substitute specification is filed, it must be submitted with markings (e.g., underlining and strikethrough) showing all the changes relative to the immediate prior version of the specification of record, it must be accompanied by a statement that the substitute specification includes no new matter, and it must be accompanied by a clean version without markings.

No change in the drawing may be made except by permission of the Office. Changes in the construction shown in any drawing may be made only by submitting replacement drawing sheets, each of which must be labeled "Replacement Sheet" in its top margin if it replaces an existing drawing sheet. Any replacement sheet of drawings must include all of the figures appearing on the immediate prior version of the sheet, even if only one figure is amended. Any new sheet of drawings containing an additional figure must be labeled in the top margin as "New Sheet." All changes to the drawings must be explained, in detail, in either the drawing amendment or remarks section of the amendment paper.

Amendments to the claims are to be made by presenting all of the claims in a claim listing that replaces all prior versions of the claims in the application. In the claim listing, the status of every claim must be indicated after its claim number after using one of the seven parenthetical expressions set forth in 37 CFR 1.121(c). "Currently amended" claims must be submitted with markings (e.g., underlining and strikethrough). All pending claims not being currently amended must be presented in the claim listing in clean version without any markings (e.g., underlining and strikethrough).

The original numbering of the claims must be preserved throughout the prosecution. When claims are canceled, the remaining claims must not be renumbered. When claims are added by amendment or substituted for canceled claims, they must be numbered by the applicant consecutively beginning with the number next following the highest numbered claim previously presented. When the application is ready for allowance, the examiner, if necessary, will renumber the claims consecutively in the order in which they appear or in such order as may have been requested by applicant.

## Time for Reply and Abandonment

The reply of an applicant to an action by the Office must be made within a prescribed time limit. The maximum period for reply is set at six months by the statute (35 U.S.C. 133), which also provides that the Director may shorten the time for reply to not less than 30 days. The usual period for reply to an Office action is three months. A shortened time for reply may be extended up to the maximum six-month period. An extension of time fee is normally required to be paid if the reply period is extended. The amount of the fee is dependent upon the length of the extension. Extensions of time are generally not available after an application has been allowed. If no reply is received within the time period, the application is considered as abandoned and no longer pending. However, if it can be shown that the failure to prosecute was unavoidable or unintentional, the application may be revived upon request to and approval by the Director. The revival requires a petition to the Director, and a fee for the petition, which must be filed without delay. The proper reply must also accompany the petition if it has not yet been filed.

## Dealing with Office Actions

The non-final office action summary indicated that the 1-10 claims in the application were pending, and were also being rejected. It was heart breaking that after several months of waiting my efforts have appeared to be for naught.

The detailed actions were initially depressing as well, showing misuse of punctuation marks, lack of knowledge on my part on the use of antecedents, and lack of clarity on the written claims. I seriously considered abandoning my application for the patent but the invaluable book "Patent It Yourself" says that it is not uncommon to have all claims rejected at the start of prosecution of an application for a patent.

Buoying my resolve to pursue the application was the fact that the application was being held in pending status and the 3-month period within which to respond to the concerns of the USPTO examiner. In short, I was being given a chance to refute the examiner's concerns.

Tabulated claims changes:

| Original Claims | As Amended by Amendment A | As Amended by Amendment B |
|---|---|---|
| 1. a digital camera lens guard and use extender for adding deviant picture taking capacities to a digital camera, and protecting said camera's lens and electronics, comprising | 11. A digital camera lens guard and use extender for adding deviant picture-taking capacities to a digital camera, and protecting said camera's lens, barrel, and electronics, comprising | 11. A digital camera lens guard and use extender for adding deviant picture-taking capacities to a digital camera, and protecting said camera's lens, barrel, and electronics, comprising |
| a. a camera mount assembly having a circular base and a cylindrical ring of sufficient size and length, for providing enclosure and track for said camera's barrel | a. a camera mount assembly having a circular base and a cylindrical ring of sufficient size and length and a plurality of locking means, for allowing unobstructed movement of said digital camera's barrel | a. a camera mount assembly having a circular base and a cylindrical ring of sufficient size and length and a plurality of locking means, for allowing unobstructed movement of said digital camera's barrel |
| b. a main housing assembly having a cylindrical ring of sufficient size and length with a plurality of retaining screws and adapter rings, for accepting and securing matching camera filters and sighting devices | b. a main housing assembly having a primary ring of sufficient size and length with a plurality of retaining screws and embed locations for adapter rings on each end, for union with camera filters and sighting devices | b. a main housing assembly having a primary ring of sufficient size and length with a plurality of retaining screws and embed locations for adapter rings on each end, for union with camera filters and sighting devices |
| 2. the digital camera lens guard and use extender of claim 1, wherein said camera mount assembly having a locking method, for engaging said circular base and said cylindrical ring together. | Cancelled | Cancelled |
| 3. the digital camera lens guard and use extender of claim 2 wherein said locking method is of slip type. | Cancelled | Cancelled |
| ... | ... | ... |
| ... wherein ... below the ... of said cylindrical ring | ... location is a circular ... ngth to one-fourth inch at each end of said primary ring of said main housing assembly. | ... embed location is ... carved apparatuses, ... ngth to one-fourth inch at each end of said primary ring of said main housing assembly. |
| 8. the digital camera lens guard and use extender of claim 6 wherein said embed location is on the surface at the end of said cylindrical ring. | Cancelled | Cancelled |
| 9. the digital camera lens guard and use extender of claim 8 wherein said embed location is in-between, a cut-off portion of said cylindrical ring and said cylindrical ring itself. | Cancelled | Cancelled |
| 10. the digital camera lens guard and use extender of claim 1 wherein said cylindrical ring of said main housing assembly being inserted with a coupler ring, for securing the eyepiece of a sighting device, whereby the digital camera lens guard and use extender adds picture taking capacities that are not normally common with a digital camera, provides protection for the camera's lens and electronics, facilitates the use of camera filters, adapter rings, and sighting devices, enables the digital camera to take pictures normally, enables the camera to take pictures of images formed at the eyepiece of a sighting device, and prepares itself for future changes and improvements. | 14. The digital camera lens guard and use extender of claim 11 wherein said primary ring of said main housing assembly being inserted with a coupler ring, for securing an eyepiece of a sighting device, whereby the digital camera lens guard and use extender adds picture-taking capacities to said digital camera, besides that of point-and-shoot method, provides protection for said digital camera's lens, barrel, and electronics, facilitates usage of camera filters, adapter rings, and sighting devices, enables said digital camera to take pictures of images formed at said eyepiece of a sighting device, and allows said digital camera to accept new picture-taking capacities that are introduced in the form of adapter rings. | 14. The digital camera lens guard and use extender of claim 11 wherein said primary ring of said main housing assembly being inserted with a coupler ring, for securing an eyepiece of a sighting device, whereby the digital camera lens guard and use extender adds picture-taking capacities to said digital camera, besides that of point-and-shoot method, provides protection for said digital camera's lens, barrel, and electronics, facilitates usage of camera filters, adapter rings, and sighting devices, enables said digital camera to take pictures of images formed at said eyepiece of a sighting device, and allows said digital camera to accept new picture-taking capacities that are introduced in the form of adapter rings. |

I followed the steps for amending the claims which necessitated responding to each and every item in the detailed action form. The detailed actions and my responses are cross-indexed for easy reference in this book.

I cancelled six claims from the non-final action form but cancellations were not easy either owing to my not wanting to accidentally delete words that were in good standing already, or add ones that might create questionable usage. I did not cancel the six claims just for the sake of cancellation. I strived to make the remaining claims tightly knit so that I only used one capital, one period, and no dashes, quotes, parentheses, trademarks, or abbreviations.

Creating a table helped me in the preceding regard; I believed it did the examiner too. Nowhere did I see a requirement for a table, but anything that would facilitate the examiner's understanding of the claims would also help my cause, I assumed.

As it turned out, with fewer claims, my invention became clearer and unmistakable. In the final action form, one more claim was cancelled leaving a total of three claims to support the invention. With the three claims the invention became even broader and can be easily built and understood by a person having ordinary skill in the art (PHOSITA).

Armed with the reduced claims and following the format in "Patent It Yourself", I sent the amendments to USPTO. The result is the grant of patent for the invention "Digital Camera Lens Guard and Use Extender."

# Non-Final Office Action

## Office Action Summary

3 months to reply

Action is non-final

Claims 1-10 pending

Claims 1-10 rejected

| Office Action Summary | Application No. 14/028,786 | Applicant(s) LIQUE, ROY | | |
|---|---|---|---|---|
| | Examiner Christopher Mahoney | Art Unit 2852 | AIA (First Inventor to File) Status Yes | |

*-- The MAILING DATE of this communication appears on the cover sheet with the correspondence address --*

**Period for Reply**

A SHORTENED STATUTORY PERIOD FOR REPLY IS SET TO EXPIRE *3* MONTHS FROM THE MAILING DATE OF THIS COMMUNICATION.

- Extensions of time may be available under the provisions of 37 CFR 1.136(a). In no event, however, may a reply be timely filed after SIX (6) MONTHS from the mailing date of this communication.
- If NO period for reply is specified above, the maximum statutory period will apply and will expire SIX (6) MONTHS from the mailing date of this communication.
- Failure to reply within the set or extended period for reply will, by statute, cause the application to become ABANDONED (35 U.S.C. § 133).
- Any reply received by the Office later than three months after the mailing date of this communication, even if timely filed, may reduce any earned patent term adjustment. See 37 CFR 1.704(b).

**Status**

1) ☐ Responsive to communication(s) filed on _____.
   ☐ A declaration(s)/affidavit(s) under **37 CFR 1.130(b)** was/were filed on _____.

2a) ☐ This action is FINAL.    2b) ☒ This action is non-final.

3) ☐ An election was made by the applicant in response to a restriction requirement set forth during the interview on _____; the restriction requirement and election have been incorporated into this action.

4) ☐ Since this application is in condition for allowance except for formal matters, prosecution as to the merits is closed in accordance with the practice under *Ex parte Quayle*, 1935 C.D. 11, 453 O.G. 213.

**Disposition of Claims***

5) ☒ Claim(s) *1-10* is/are pending in the application.
   5a) Of the above claim(s) _____ is/are withdrawn from consideration.

6) ☐ Claim(s) _____ is/are allowed.

7) ☒ Claim(s) *1-10* is/are rejected.

8) ☐ Claim(s) _____ is/are objected to.

9) ☐ Claim(s) _____ are subject to restriction and/or election requirement.

* If any claims have been determined <u>allowable</u>, you may be eligible to benefit from the **Patent Prosecution Highway** program at a participating intellectual property office for the corresponding application. For more information, please see http://www.uspto.gov/patents/init_events/pph/index.jsp or send an inquiry to PPHfeedback@uspto.gov.

**Application Papers**

10) ☐ The specification is objected to by the Examiner.

11) ☐ The drawing(s) filed on _____ is/are: a) ☐ accepted or b) ☐ objected to by the Examiner.
    Applicant may not request that any objection to the drawing(s) be held in abeyance. See 37 CFR 1.85(a).
    Replacement drawing sheet(s) including the correction is required if the drawing(s) is objected to. See 37 CFR 1.121(d).

**Priority under 35 U.S.C. § 119**

12) ☐ Acknowledgment is made of a claim for foreign priority under 35 U.S.C. § 119(a)-(d) or (f).
    **Certified copies:**
    a) ☐ All    b) ☐ Some**    c) ☐ None of the:
    1. ☐ Certified copies of the priority documents have been received.
    2. ☐ Certified copies of the priority documents have been received in Application No. _____.
    3. ☐ Copies of the certified copies of the priority documents have been received in this National Stage application from the International Bureau (PCT Rule 17.2(a)).
    ** See the attached detailed Office action for a list of the certified copies not received.

**Attachment(s)**

1) ☒ Notice of References Cited (PTO-892)

2) ☒ Information Disclosure Statement(s) (PTO/SB/08a and/or PTO/SB/08b)
   Paper No(s)/Mail Date *Oct. 22, 2013*

3) ☐ Interview Summary (PTO-413)
   Paper No(s)/Mail Date _____

4) ☐ Other: _____

U.S. Patent and Trademark Office
PTOL-326 (Rev. 11-13)    Office Action Summary    Part of Paper No./Mail Date 35

# Detailed Action - Page 1

Detailed Action 1

Detailed Action 2

Detailed Action 3

Application/Control Number: 14/028,786    Page 2
Art Unit: 2852

The present application, filed on or after March 16, 2013, is being examined under the

first inventor to file provisions of the AIA.

## DETAILED ACTION

### Claim Objections

Claims 1-9 are objected to because of the following informalities:

Claims 1-9 all end with a comma instead of a period.

The examiner believes "locations" should be singular "location" in claim 6

Appropriate correction is required.

### Claim Rejections - 35 USC § 112

The following is a quotation of the first paragraph of 35 U.S.C. 112(a):

(a) IN GENERAL.— The specification shall contain a written description of the invention, and of the manner and process of making and using it, in such full, clear, concise, and exact terms as to enable any person skilled in the art to which it pertains, or with which it is most nearly connected, to make and use the same, and shall set forth the best mode contemplated by the inventor or joint inventor of carrying out the invention.

The following is a quotation of the first paragraph of pre-AIA 35 U.S.C. 112:

The specification shall contain a written description of the invention, and of the manner and process of making and using it, in such full, clear, concise, and exact terms as to enable any person skilled in the art to which it pertains, or with which it is most nearly connected, to make and use the same, and shall set forth the best mode contemplated by the inventor of carrying out his invention.

Claim 9 is rejected under 35 U.S.C. 112(a) or 35 U.S.C. 112 (pre-AIA), first paragraph,

as failing to comply with the written description requirement. The claim(s) contains subject

matter which was not described in the specification in such a way as to reasonably convey to one

skilled in the relevant art that the inventor or a joint inventor, or for pre-AIA the inventor(s), at

# Detailed Action - Page 2

Application/Control Number: 14/028,786        Page 3
Art Unit: 2852

the time the application was filed, had possession of the claimed invention. The specification

does not describe the embed location between a cut off portion and the cylindrical ring.

The following is a quotation of 35 U.S.C. 112(b):
(b) CONCLUSION. — The specification shall conclude with one or more claims particularly pointing
out and distinctly claiming the subject matter which the inventor or a joint inventor regards as the
invention.

The following is a quotation of 35 U.S.C. 112 (pre-AIA), second paragraph:
The specification shall conclude with one or more claims particularly pointing out and distinctly
claiming the subject matter which the applicant regards as his invention.

Claims 1-10 are rejected under 35 U.S.C. 112(b) or 35 U.S.C. 112 (pre-AIA), second

paragraph, as being indefinite for failing to particularly point out and distinctly claim the subject

matter which the inventor or a joint inventor, or for pre-AIA the applicant regards as the

invention.

Claim 1 recites a cylindrical ring in subparagraph a and then recites a cylindrical ring in

subparagraph b. It is unclear if the same or different cylindrical rings are being claimed/referred

to.

There is a lack of antecedent basis for "said camera's barrel" as recited in claim 1.

There is a lack of antecedent basis for "the surface" and "the end" recited in claim 7.

There is a lack of antecedent basis for "the surface" and "the end" recited in claim 8.

It is unclear what is embedded in claim 8 if the embed location is on the surface at the

end.

It is unclear which cylindrical ring is being referred to in claim 9.

It is unclear what is meant by the "embed location is in-between a cut off portion of said

cylindrical ring and said cylindrical ring itself" as recited in claim 9.

## Detailed Action - Page 3

Application/Control Number: 14/028,786                                    Page 4
Art Unit: 2852

The language "prepares itself for future changes and improvements" in claim 10 is unclear and indefinite.

It is unclear what is meant by "not normally common" as recited in claim 10.

There is a lack of antecedent basis for "the eyepiece" recited in claim 10.

### Claim Rejections - 35 USC § 102

The following is a quotation of the appropriate paragraphs of 35 U.S.C. 102 that form the basis for the rejections under this section made in this Office action:

A person shall be entitled to a patent unless –

(a)(1) the claimed invention was patented, described in a printed publication, or in public use, on sale or otherwise available to the public before the effective filing date of the claimed invention.

Claims 1-3, 5-7 and 9-10 are rejected under 35 U.S.C. 102(a)(1) as being anticipated by Nelson et al. (U.S. Patent No. 3,752,569). Nelson teaches a camera mount assembly having a circular base and a cylindrical ring 90 of sufficient size and length, for providing enclosure and track for said camera's barrel, a main housing assembly having a cylindrical ring 100 of sufficient size and length with a plurality of retaining screws102/106 and adapter rings 114, for accepting and securing matching camera filters and sighting devices.

The claim language "for providing enclosure and track" and "for accepting and securing" are considered intended uses and not necessarily given patentable weight.

The circular base and the cylindrical ring have a locking method for engaging said circular base and said cylindrical ring together as shown in figure 4.

# Detailed Action - Page 4

Detailed Action 19

Detailed Action 20

Detailed Action 21

Regarding claim 6, the cylindrical ring of the main housing assembly has more than one embed location. One embed location is in the front (right side in figure 4) and one embed location is in the back (left side in figure 4).

### *Claim Rejections - 35 USC § 103*

The following is a quotation of 35 U.S.C. 103 which forms the basis for all obviousness rejections set forth in this Office action:

A patent for a claimed invention may not be obtained, notwithstanding that the claimed invention is not identically disclosed as set forth in section 102 of this title, if the differences between the claimed invention and the prior art are such that the claimed invention as a whole would have been obvious before the effective filing date of the claimed invention to a person having ordinary skill in the art to which the claimed invention pertains. Patentability shall not be negated by the manner in which the invention was made.

Claim 4 is rejected under 35 U.S.C. 103 as being unpatentable over Nelson et al. (U.S. Patent No. 3,752,569) in view of Spencer (U.S. Patent No. 7088918). Nelson teaches the salient features of the claimed invention except for a screw type locking method. Spencer teaches in figures 3 and 4 as well as col. 3, lines 1—22 that is was known to provide a threaded connector for lens adapters. It would have been obvious to one of ordinary skill in the art at the time the invention was made to utilize the features of Spencer for the purpose of quick connections.

Claim 8 is rejected under 35 U.S.C. 103 as being unpatentable over Nelson et al. (U.S. Patent No. 3,752,569) in view of MacKay (U.S. Patent No. 5208624). Nelson teaches the salient features of the claimed invention except for the embed location on the surface. MacKay teaches that it was known to embed the adapter on the surface via magnets. It would have been obvious to one of ordinary skill in the art at the time the invention was made to utilize the features of MacKay for the purpose of quick disconnects.

**Detailed Action - Page 5**

Application/Control Number: 14/028,786    Page 6
Art Unit: 2852

### *Conclusion*

The prior art made of record and not relied upon is considered pertinent to applicant's disclosure.

Any inquiry concerning this communication or earlier communications from the examiner should be directed to Christopher Mahoney whose telephone number is (571)272-2122. The examiner can normally be reached on 8:30AM-5PM, Monday-Thursday.

If attempts to reach the examiner by telephone are unsuccessful, the examiner's supervisor, Clayton Laballe can be reached on (571) 272-1594. The fax phone number for the organization where this application or proceeding is assigned is 571-273-8300.

Information regarding the status of an application may be obtained from the Patent Application Information Retrieval (PAIR) system. Status information for published applications may be obtained from either Private PAIR or Public PAIR. Status information for unpublished applications is available through Private PAIR only. For more information about the PAIR system, see http://pair-direct.uspto.gov. Should you have questions on access to the Private PAIR system, contact the Electronic Business Center (EBC) at 866-217-9197 (toll-free). If you would like assistance from a USPTO Customer Service Representative or access to the automated information system, call 800-786-9199 (IN USA OR CANADA) or 571-272-1000.

/Christopher Mahoney/
Primary Examiner, Art Unit 2852

# Notice of References Cited

| | | Application/Control No. 14/028,786 | | Applicant(s)/Patent Under Reexamination LIQUE, ROY | | |
|---|---|---|---|---|---|---|
| **Notice of References Cited** | | Examiner Christopher Mahoney | | Art Unit 2852 | | Page 1 of 1 |

**U.S. PATENT DOCUMENTS**

| * | | Document Number Country Code-Number-Kind Code | Date MM-YYYY | Name | Classification |
|---|---|---|---|---|---|
| * | A | US-2,953,970 A | 09-1960 | MAYNARD CHARLES A | 359/611 |
| * | B | US-3,752,569 A | 08-1973 | Nelson et al. | 352/139 |
| * | C | US-5,208,624 A | 05-1993 | MacKay, Michael T. | 396/544 |
| * | D | US-5,640,630 A | 06-1997 | Hattan, Mark | 396/342 |
| * | E | US-2004/0252987 A1 | 12-2004 | Kobayashi, Fumio | 396/006 |
| * | F | US-7,068,918 B1 | 06-2006 | Spencer, Randolph | 396/432 |
| * | G | US-7,450,325 B2 | 11-2008 | Yamashita et al. | 359/630 |
| * | H | US-2009/0109556 A1 | 04-2009 | Schaefer, Glenn F. | 359/827 |
| * | I | US-7,813,632 B2 | 10-2010 | Laganas et al. | 396/71 |
| * | J | US-2013/0094101 A1 | 04-2013 | Oguchi, Yasunari | 359/745 |
| * | K | US-2013/0129335 A1 | 05-2013 | Gainer, Robert | 396/144 |
| * | L | US-2013/0230309 A1 | 09-2013 | Porter et al. | 396/432 |
| * | M | US-8,593,742 B2 | 11-2013 | Takahashi, Kazunori | 359/619 |

**FOREIGN PATENT DOCUMENTS**

| * | | Document Number Country Code-Number-Kind Code | Date MM-YYYY | Country | Name | Classification |
|---|---|---|---|---|---|---|
| | N | | | | | |
| | O | | | | | |
| | P | | | | | |
| | Q | | | | | |
| | R | | | | | |
| | S | | | | | |
| | T | | | | | |

**NON-PATENT DOCUMENTS**

| * | | Include as applicable: Author, Title Date, Publisher, Edition or Volume, Pertinent Pages) |
|---|---|---|
| | U | |
| | V | |
| | W | |
| | X | |

*A copy of this reference is not being furnished with this Office action. (See MPEP § 707.05(a).)
Dates in MM-YYYY format are publication dates. Classifications may be US or foreign.

U.S. Patent and Trademark Office
PTO-892 (Rev. 01-2001)        **Notice of References Cited**        Part of Paper No. 35

# East Search History – Page 1

EAST Search History

EAST Search History

EAST Search History (Prior Art)

| Ref # | Hits | Search Query | DBs | Default Operator | Plurals | Time Stamp |
|---|---|---|---|---|---|---|
| S1 | 49 | (("20150077626") or ("8284254") or ("20050185089") or ("4156269") or ("4912494") or ("5053798") or ("5640630") or ("5708868") or ("5794090") or ("6694096") or ("7489455") or ("7532812") or ("7880795") or ("7880795") or ("8126325") or ("8698894") or ("20070058963") or ("20080062539") or ("20080136941") or ("20090196591") or ("20100290767") or ("20120242881") or ("20140218534") or ("5475441") or ("6449364") or ("6603509") or ("6829624") or ("20010036293") or ("20010043717") or ("20010043718") or ("20020112112") or ("4288814") or ("4876563") or ("5009499") or ("5287644") or ("6688870") or ("7149418") or ("8358928") or ("20020066407") or ("20060067657") or ("20110249965") or ("20140221749") or ("5923363") or ("6502787") or ("6854226") or ("20020060737") or ("20040136388") or ("20050271266") or ("20070039030") or ("20120236154")).PN. | US-PGPUB; USPAT; USOCR | OR | OFF | 2015/04/20 13:51 |
| S2 | 18 | ("3614196" | "3840883" | "3909167" | "4137540" | "4285706" | "4384767" | "4533212" | "5126881" | "6243540" | "7031081" | "7161749" | "7358131" | "7386229" | "7813639" | "7982981" | "8045277" | "8054545" | "8111964").PN. | US-PGPUB; USPAT; USOCR | OR | OFF | 2015/04/20 13:51 |
| S3 | 4486 | g03b17/14.cpc. | US-PGPUB; USPAT; USOCR; FPRS; EPO; JPO; DERWENT; IBM_TDB | OR | OFF | 2015/04/23 16:31 |
| S4 | 2356 | g03b17/12.cpc. | US-PGPUB; USPAT; USOCR; FPRS; EPO; JPO; DERWENT; IBM_TDB | OR | OFF | 2015/04/23 16:32 |
| S5 | 495 | g03b17/565.cpc. | US-PGPUB; USPAT; USOCR; FPRS; EPO; JPO; | OR | OFF | 2015/04/23 16:32 |

## East Search History – Page 2

EAST Search History

| | | | DERWENT; IBM_TDB | | | |
|---|---|---|---|---|---|---|
| S6 | 3690 | g03b17/56.cpc. | US-PGPUB; USPAT; USOCR; FPRS; EPO; JPO; DERWENT; IBM_TDB | OR | OFF | 2015/04/23 16:32 |
| S7 | 15170 | h04n5/2254.cpc. | US-PGPUB; USPAT; USOCR; FPRS; EPO; JPO; DERWENT; IBM_TDB | OR | OFF | 2015/04/23 16:35 |
| S8 | 393 | (S3 S4) and (S5 S6) | US-PGPUB; USPAT; USOCR; FPRS; EPO; JPO; DERWENT; IBM_TDB | OR | OFF | 2015/04/23 16:35 |
| S9 | 326 | (S3 ) and (S5 S6) | US-PGPUB; USPAT; USOCR; FPRS; EPO; JPO; DERWENT; IBM_TDB | OR | OFF | 2015/04/23 16:35 |
| S10 | 221 | (S3 ) and (S5) | US-PGPUB; USPAT; USOCR; FPRS; EPO; JPO; DERWENT; IBM_TDB | OR | OFF | 2015/04/23 16:35 |
| S11 | 233 | (S3 S4) and (S5) | US-PGPUB; USPAT; USOCR; FPRS; EPO; JPO; DERWENT; IBM_TDB | OR | OFF | 2015/04/23 16:35 |
| S13 | 1 | (13/081838).AP P. | USPAT; USOCR | OR | OFF | 2015/04/23 16:36 |
| S14 | 5 | "2011072604".pn. | US-PGPUB; USPAT; USOCR; FPRS; EPO; JPO; DERWENT; IBM_TDB | OR | OFF | 2015/04/23 17:40 |
| S15 | 4486 | g03b17/14.cpc. | US-PGPUB; USPAT; USOCR; FPRS; EPO; JPO; DERWENT; IBM_TDB | OR | OFF | 2015/04/23 17:44 |

file:///C/Users/cmahoney/Documents/e-Red%20Folder/14028786/EASTSearchHistory.14028786_AccessibleVersion.htm[5/4/2015 1:09:35 PM]

# East Search History – Page 3

EAST Search History

| S16 | 2355 | g03b17/12.cpc. | US-PGPUB; USPAT; USOCR; FPRS; EPO; JPO; DERWENT; IBM_TDB | OR | OFF | 2015/04/23 17:44 |
|---|---|---|---|---|---|---|
| S17 | 495 | g03b17/565.cpc. | US-PGPUB; USPAT; USOCR; FPRS; EPO; JPO; DERWENT; IBM_TDB | OR | OFF | 2015/04/23 17:44 |
| S18 | 3690 | g03b17/56.cpc. | US-PGPUB; USPAT; USOCR; FPRS; EPO; JPO; DERWENT; IBM_TDB | OR | OFF | 2015/04/23 17:44 |
| S19 | 393 | (S15 S16) and (S17 S18) | US-PGPUB; USPAT; USOCR; FPRS; EPO; JPO; DERWENT; IBM_TDB | OR | OFF | 2015/04/23 17:44 |
| S20 | 221 | (S15 ) and (S17) | US-PGPUB; USPAT; USOCR; FPRS; EPO; JPO; DERWENT; IBM_TDB | OR | OFF | 2015/04/23 17:44 |
| S21 | 172 | S19 not S20 | US-PGPUB; USPAT; USOCR; FPRS; EPO; JPO; DERWENT; IBM_TDB | OR | OFF | 2015/04/23 17:44 |
| S22 | 158 | (396/530).CCLS. | US-PGPUB; USPAT; USOCR; FPRS; EPO; JPO; DERWENT; IBM_TDB | OR | OFF | 2015/04/23 18:15 |
| S23 | 1371 | (396/529).CCLS. | US-PGPUB; USPAT; USOCR; FPRS; EPO; JPO; DERWENT; IBM_TDB | OR | OFF | 2015/04/23 18:15 |
| S24 | 207 | (396/71).CCLS. | US-PGPUB; USPAT; USOCR; FPRS; EPO; JPO; | OR | OFF | 2015/04/23 18:15 |

file:///C/Users/cmahoney/Documents/e-Red%20Folder/14028786/EAST Search History.14028786_AccessibleVersion.htm[5/4/2015 1:09:35 PM]

# East Search History – Page 4

EAST Search History

| | | | | DERWENT; IBM_TDB | | | |
|---|---|---|---|---|---|---|---|
| S25 | 984 | (396/544).CCLS. | | US-PGPUB; USPAT; USOCR; FPRS; EPO; JPO; DERWENT; IBM_TDB | OR | OFF | 2015/04/23 18:15 |
| S26 | 198 | (359/611).CCLS. | | USPAT; USOCR | OR | OFF | 2015/04/26 14:01 |
| S27 | 278 | (359/611).CCLS. | | US-PGPUB; USPAT; USOCR; FPRS; EPO; JPO; DERWENT; IBM_TDB | OR | OFF | 2015/04/26 14:01 |
| S28 | 117 | (359/612).CCLS. | | US-PGPUB; USPAT; USOCR; FPRS; EPO; JPO; DERWENT; IBM_TDB | OR | OFF | 2015/04/26 14:01 |
| S29 | 6182 | ((359/811) or (359/819) or (359/822)).CCLS. | | US-PGPUB; USPAT; USOCR; FPRS; EPO; JPO; DERWENT; IBM_TDB | OR | OFF | 2015/04/26 14:03 |
| S30 | 494 | (359/826).CCLS. | | US-PGPUB; USPAT; USOCR; FPRS; EPO; JPO; DERWENT; IBM_TDB | OR | OFF | 2015/04/26 14:03 |
| S31 | 2873 | (359/823).CCLS. | | US-PGPUB; USPAT; USOCR; FPRS; EPO; JPO; DERWENT; IBM_TDB | OR | OFF | 2015/04/26 14:04 |
| S32 | 318 | (359/830).CCLS. | | US-PGPUB; USPAT; USOCR; FPRS; EPO; JPO; DERWENT; IBM_TDB | OR | OFF | 2015/04/26 14:04 |
| S33 | 24 | (S27 S28) and (S29 S30 S31 S32 ) | | US-PGPUB; USPAT; USOCR; FPRS; EPO; JPO; DERWENT; IBM_TDB | OR | OFF | 2015/04/26 14:04 |

file:///C:/Users/cmahoney/Documents/e-Red%20Folder/14028786/EASTSearchHistory.14028786_AccessibleVersion.htm[5/4/2015 1:09:35 PM]

## East Search History – Page 5

EAST Search History

| S34 | 2104 | (359/811).CCLS. | US-PGPUB; USPAT; USOCR; FPRS; EPO; JPO; DERWENT; IBM_TDB | OR | OFF | 2015/04/28 14:08 |
|-----|------|-----------------|---------|----|-----|-----------|
| S35 | 3855 | (359/619).CCLS. | US-PGPUB; USPAT; USOCR; FPRS; EPO; JPO; DERWENT; IBM_TDB | OR | OFF | 2015/04/28 14:08 |
| S36 | 329969 | (set setting lock$4) near screw | US-PGPUB; USPAT; USOCR; FPRS; EPO; JPO; DERWENT; IBM_TDB | OR | ON | 2015/04/28 14:09 |
| S37 | 337 | S29 and S36 | US-PGPUB; USPAT; USOCR; FPRS; EPO; JPO; DERWENT; IBM_TDB | OR | ON | 2015/04/28 14:09 |
| S38 | 21 | S37 and "396"/$.ccls. | US-PGPUB; USPAT; USOCR; FPRS; EPO; JPO; DERWENT; IBM_TDB | OR | ON | 2015/04/28 14:09 |
| S39 | 1 | ("6593742").PN. | USPAT; USOCR | OR | OFF | 2015/04/28 14:25 |
| S40 | 6182 | ((359/811) or (359/619) or (359/822)).CCLS. | US-PGPUB; USPAT; USOCR; FPRS; EPO; JPO; DERWENT; IBM_TDB | OR | OFF | 2015/04/28 22:40 |
| S41 | 494 | (359/826).CCLS. | US-PGPUB; USPAT; USOCR; FPRS; EPO; JPO; DERWENT; IBM_TDB | OR | OFF | 2015/04/28 22:40 |
| S42 | 2873 | (359/823).CCLS. | US-PGPUB; USPAT; USOCR; FPRS; EPO; JPO; DERWENT; IBM_TDB | OR | OFF | 2015/04/28 22:40 |
| S43 | 2104 | (359/811).CCLS. | US-PGPUB; USPAT; USOCR | OR | OFF | 2015/04/28 22:40 |

file:///C/Users/cmahoney/Documents/e-Red%20Folder/14028786/EASTSearchHistory.14028786_AccessibleVersion.htm[5/4/2015 1:09:35 PM]

# East Search History – Page 6

EAST Search History

| | | | FPRS; EPO; JPO; DERWENT; IBM_TDB | | | |
|---|---|---|---|---|---|---|
| S44 | 3855 | (359/819).CCLS. | US-PGPUB; USPAT; USOCR; FPRS; EPO; JPO; DERWENT; IBM_TDB | OR | OFF | 2015/04/28 22:40 |
| S45 | 117 | (359/612).CCLS. | US-PGPUB; USPAT; USOCR; FPRS; EPO; JPO; DERWENT; IBM_TDB | OR | OFF | 2015/04/28 22:41 |
| S46 | 278 | (359/611).CCLS. | US-PGPUB; USPAT; USOCR; FPRS; EPO; JPO; DERWENT; IBM_TDB | OR | OFF | 2015/04/28 22:41 |
| S47 | 9191 | S40 S41 S42 S43 S44 S45 S46 | US-PGPUB; USPAT; USOCR; FPRS; EPO; JPO; DERWENT; IBM_TDB | OR | OFF | 2015/04/28 22:41 |
| S48 | 107 | bushings and S47 | US-PGPUB; USPAT; USOCR; FPRS; EPO; JPO; DERWENT; IBM_TDB | OR | ON | 2015/04/28 22:41 |
| S49 | 329969 | (set setting lock$4) near screw | US-PGPUB; USPAT; USOCR; FPRS; EPO; JPO; DERWENT; IBM_TDB | OR | ON | 2015/04/28 22:42 |
| S50 | 33 | S48 and S49 | US-PGPUB; USPAT; USOCR; FPRS; EPO; JPO; DERWENT; IBM_TDB | OR | ON | 2015/04/28 22:42 |
| S51 | 1 | ("20150077626").PN. | US-PGPUB; USPAT; USOCR | OR | OFF | 2015/05/04 12:14 |

**EAST Search History (Interference)**

< This search history is empty>

file:///C:/Users/cmahoney/Documents/e-Red%20Folder/14028786/EASTSearchHistory.14028786_AccessibleVersion.htm[5/4/2015 1:09:35 PM]

224

**East Search History – Page 7**

EAST Search History

5/ 4/ 2015 1:09:30 PM
C:\ Users\ cmahoney\ Documents\ EAST\ Workspaces\ 14028786.wsp

file:///C/Users/cmahoney/Documents/b-Re/P%20Folder/14028786/EAST Search History.14028786_AccessibleVersion.htm[5/4/2015 1:0:35 PM]

**BIB Data Sheet**

Page 1 of 1

UNITED STATES PATENT AND TRADEMARK OFFICE

UNITED STATES DEPARTMENT OF COMMERCE
United States Patent and Trademark Office
Address: COMMISSIONER FOR PATENTS
P.O. Box 1450
Alexandria, Virginia 22313-1450
www.uspto.gov

## BIB DATA SHEET

CONFIRMATION NO. 6716

| SERIAL NUMBER | FILING or 371(c) DATE | CLASS | GROUP ART UNIT | ATTORNEY DOCKET NO. |
|---|---|---|---|---|
| 14/028,786 | 09/17/2013 RULE | 396 | 2852 | Camera Coup |

**APPLICANTS**

**INVENTORS**
  Roy Lique, Walnut, CA;

** CONTINUING DATA ***********************

** FOREIGN APPLICATIONS ***********************

** IF REQUIRED, FOREIGN FILING LICENSE GRANTED ** ** MICRO ENTITY **
  10/01/2013

| Foreign Priority claimed ☐ Yes ☑ No | | | STATE OR COUNTRY | SHEETS DRAWINGS | TOTAL CLAIMS | INDEPENDENT CLAIMS |
|---|---|---|---|---|---|---|
| 35 USC 119(a-d) conditions met ☐ Yes ☑ No | ☐ Met after Allowance | | CA | 7 | 10 | 1 |
| Verified and Acknowledged  CHRIST E MAHONEY Examiner's Signature  Initials | | | | | | |

**ADDRESS**

  Roy Lique
  1431 Tierra Cima
  Walnut, CA 91789
  UNITED STATES

**TITLE**

  Digital Camera Lens Guard and Use Extender

| FILING FEE RECEIVED 435 | FEES: Authority has been given in Paper No._____ to charge/credit DEPOSIT ACCOUNT No._____ for following: | ☐ All Fees |
|---|---|---|
| | | ☐ 1.16 Fees (Filing) |
| | | ☐ 1.17 Fees (Processing Ext. of time) |
| | | ☐ 1.18 Fees (Issue) |
| | | ☐ Other |
| | | ☐ Credit |

BIB (Rev. 05/07)

226

## Search Notes

| Search Notes | Application/Control No. | Applicant(s)/Patent Under Reexamination |
|---|---|---|
| | 14028786 | LIQUE, ROY |
| | Examiner | Art Unit |
| | CHRIST MAHONEY | 2852 |

**CPC- SEARCHED**

| Symbol | Date | Examiner |
|---|---|---|
| G03B 17/12 | 4/28/2015 | CM |
| G03B 17/14 | 4/28/2015 | CM |
| G03B 17/565 | 4/28/2015 | CM |
| H04N 5/2254 | 4/28/2015 | CM |

**CPC COMBINATION SETS - SEARCHED**

| Symbol | Date | Examiner |
|---|---|---|
| | | |

**US CLASSIFICATION SEARCHED**

| Class | Subclass | Date | Examiner |
|---|---|---|---|
| 396 | 71 | 4/28/2015 | CM |
| 396 | 544 | 4/28/2015 | CM |
| 396 | 611 | 4/28/2015 | CM |
| 396 | 612 | 4/28/2015 | CM |
| 359 | 811 | 4/28/2015 | CM |
| 359 | 819 | 4/28/2015 | CM |
| 359 | 822 | 4/28/2015 | CM |
| 359 | 826 | 4/28/2015 | CM |
| 359 | 823 | 4/28/2015 | CM |

**SEARCH NOTES**

| Search Notes | Date | Examiner |
|---|---|---|
| PLUS search | 4/03/2015 | CM |
| EAST | 4/28/2015 | CM |

**INTERFERENCE SEARCH**

| US Class/ CPC Symbol | US Subclass / CPC Group | Date | Examiner |
|---|---|---|---|
| | | | |

U.S. Patent and Trademark Office

Part of Paper No. 35

## Response - Amendment A

Amendment A - Cover Letter

Appn. Number 14/028,786    (Roy Lique)    Art Unit 2852    Amnt. A    Page **1 of 10**

**In the United States Patent and Trademark Office**

Appn. Number:      14/028,786
Appn. Filed        09/17/2013
Applicants:        Roy Lique
Title:             Digital Camera Lens Guard and Use Extender
Examiner/GAU:      Christopher Mahoney/Art Unit 2852

Walnut, July 29, 2015

**AMENDMENT A**

Mail Stop Non-Fee Amendments
Commissioner for Patents
P.O. Box 1450
Alexandria, VA 22313-1450

Sir:

In response to the Office Action Mailed May 11, 2015, please amend the above application as follows:

CLAIMS: Amendments to the claims begin on page 2 of this Amendment
REMARKS: Remarks begin on page 5 of this Amendment.

Appn. Number 14/028,786    (Roy Lique)    Art Unit 2852    Amnt. A    Page **2** of **10**

**CLAIMS**

Please amend the claims according to the status designations in the following list, which contains changes of all claims that were ever in the application, with the text of all new active claims.

CLAIMS: I claim:

1.-10. (CANCELED)

11.  (NEW) A digital camera lens guard and use extender for adding deviant picture-taking capacities to a digital camera, and protecting said camera's lens, barrel, and electronics, comprising:

   a.  a camera mount assembly having a circular base and a cylindrical ring of sufficient size and length and a plurality of locking means, for allowing unobstructed movement of said digital camera's barrel,

   b.  a main housing assembly having a primary ring of sufficient size and length with a plurality of retaining screws and embed locations for adapter rings on each end, for union with camera filters and sighting devices.

12.  (NEW) The digital camera lens guard and use extender of claim 11 wherein said cylindrical ring of said camera mount assembly being fitted with an adapter ring, for directly adding picture-taking capacities to said digital camera without using said primary ring of said main housing assembly.

13.  (NEW) The digital camera lens guard and use extender of claim 11 wherein said embed location is a circular recess carved approximately one-eighth to one-fourth inch at each said end of said primary ring of said main housing assembly.

14.  (NEW The digital camera lens guard and use extender of claim 11 wherein said primary ring of said main housing assembly being inserted with a coupler ring, for securing an eyepiece of a sighting device,

whereby the digital camera lens guard and use extender adds picture-taking capacities to said digital camera beside that of point-and-shoot method; provides protection for said digital camera's lens, barrel, and electronics; facilitates usage of camera filters, adapter rings, and sighting devices; enables said digital camera to take pictures of images formed at said eyepiece of a sighting device; and allows said digital camera to accept new picture-taking capacities that are introduced in the form of adapter rings.

Appn. Number 14/028,786    (Roy Lique)    Art Unit 2852    Amnt. A    Page **3** of **10**

**Tabulated claims changes:**

| Original Claims | Amended Claims |
|---|---|
| 1. a digital camera lens guard and use extender for adding deviant picture taking capacities to a digital camera, and protecting said camera's lens and electronics, comprising, | 11. A digital camera lens guard and use extender for adding deviant picture-taking capacities to a digital camera, and protecting said camera's lens, barrel, and electronics, comprising: |
| a. a camera mount assembly having a circular base and a cylindrical ring of sufficient size and length, for providing enclosure and track for said camera's barrel. | a. a camera mount assembly having a circular base and a cylindrical ring of sufficient size and length and a plurality of locking means, for allowing unobstructed movement of said digital camera's barrel. |
| b. a main housing assembly having a cylindrical ring of sufficient size and length with a plurality of retaining screws and adapter rings, for accepting and securing matching camera filters and sighting devices. | b. a main housing assembly having a primary ring of sufficient size and length with a plurality of retaining screws and embed locations for adapter rings on each end, for union with camera filters and sighting devices. |
| 2. the digital camera lens guard and use extender of claim 1, wherein said camera mount assembly having a locking method, for engaging said circular base and said cylindrical ring together, | Canceled |
| 3. the digital camera lens guard and use extender of claim 2 wherein said locking method is of slip type. | Canceled |
| 4. the digital camera lens guard and use extender of claim 2 wherein said locking method is of screw type. | Canceled |
| 5. the digital camera lens guard and use extender of claim 1 wherein said cylindrical ring of said camera mount assembly being fitted with an adapter ring, for securing camera filters or sighting devices, | 12. The digital camera lens guard and use extender of claim 11 wherein said cylindrical ring of said camera mount assembly being fitted with an adapter ring, for directly adding picture-taking capacities to said digital camera without using said primary ring of said main housing assembly. |
| 6. the digital camera lens guard and use extender of claim 1 wherein said cylindrical ring of said main housing assembly having more than one embed locations, for securing said adapter rings of said main housing assembly, | Canceled |
| 7. the digital camera lens guard and use extender of claim 1 wherein said embed location is below the surface at the end of said cylindrical ring. | 13. The digital camera lens guard and use extender of claim 11 wherein said embed location is a circular recess carved approximately one-eighth to one-fourth inch at each said end of said primary ring of said main housing assembly. |
| 8. the digital camera lens guard and use extender of claim 6 wherein said embed location is on the surface at the end of said cylindrical ring. | Canceled |

Appn. Number 14/028,786      (Roy Lique)      Art Unit 2852      Amnt. A      Page **4** of **10**

| | |
|---|---|
| 9. the digital camera lens guard and use extender of claim 6 wherein said embed location is in-between a cut off portion of said cylindrical ring and said cylindrical ring itself. | Canceled |
| 10. the digital camera lens guard and use extender of claim 1 wherein said cylindrical ring of said main housing assembly being inserted with a coupler ring, for securing the eyepiece of a sighting device. | 14. The digital camera lens guard and use extender of claim 11 wherein said primary ring of said main housing assembly being inserted with a coupler ring, for securing an eyepiece of a sighting device. |
| whereby the digital camera lens guard and use extender adds picture taking capacities that are not normally common with a digital camera; provides protection for the camera's lens and electronics; facilitates the use of camera filters, adapter rings, and sighting devices; enables the digital camera to take pictures normally; enables the camera to take pictures of images formed at the eyepiece of a sighting device; and prepares itself for future changes and improvements. | whereby the digital camera lens guard and use extender adds picture-taking capacities to said digital camera beside that of point-and-shoot method; provides protection for said digital camera's lens, barrel, and electronics; facilitates usage of camera filters, adapter rings, and sighting devices; enables said digital camera to take pictures of images formed at said eyepiece of a sighting device; and allows said digital camera to accept new picture-taking capacities that are introduced in the form of adapter rings. |

Response 1

Response 2

Response 3

Response 4

Response 5

Response 6

**REMARKS – General**

By the above amendment, applicant has corrected, to the best of his knowledge, the following:

### Claim objections due to the following informalities

**Claims 1-9 all end with a comma instead of a period.**
Claims 1-10 have all been canceled and replaced with claims 11-14. Claims 11-14 have now been properly punctuated.

**"locations" in claim 6 should be singular instead of plural.**
Claim 6 has been canceled. All its texts including "locations" no longer exist in the amended claims.

### Claim Rejection – 35 USC § 112

**Claim 9 is rejected under 35 U.S.C. 112(a) or 35 U.S.C. 112 (pre-AIA), first paragraph, as failing to comply with the written description requirement.**
Claims 1-10 have all been canceled and replaced with claims 11-14. Claim 9 no longer exists in the amended claims.

**Claims 1-10 are rejected under 35 Ul.S.C. 112(b) or 35 U.S.C. 112 (pre-AIA), second paragraph, as being indefinite for failing to particularly point out and distinctly claim the subject matter which the inventor or a joint inventor, or for pre-AIA the applicant regards as the invention.**
Claims 1-10 have all been canceled and replaced with claims 11-14. Responses to bases for rejections follow.

**Claim 1 recites a cylindrical ring in subparagraph a and then recites a cylindrical ring in subparagraph b. It is unclear if the same or different cylindrical are being claimed/referred to.**
Claim 1 is now claim 11 in the amended claims. Subparagraph a still recites a cylindrical ring; subparagraph b now recites a primary ring. The subjects of both recitations are now different and definite.

**There is a lack of antecedent basis for "said camera's barrel" as recited in claim 1.**
Claim 1 is now claim 11 in the amended claims. This basis for rejection has now been corrected by adding "barrel" to paragraph in claim 11. Antecedent basis for "said camera's barrel" has been established.

Response 7

Response 8

Response 9

Response 10

Response 11

Response 12

Response 13

Response 14

Appn. Number 14/028,786     (Roy Lique)     Art Unit 2852     Amnt. A     Page 6 of 10

**There is lack of antecedent basis for "the surface" and "the end" recited in claim 7.** Claim 7 is now claim 13 in the amended claims. Claim 13 has corrected the lack antecedent basis for "the surface" by replacing the language "is below the surface" with the language "is a circular recess carved approximately one-eighth to one-fourth inch".

Claim 13 has corrected the lack of antecedent basis for "the end" by replacing the language "the end of said cylindrical ring" with the language "each said end of said primary ring of said main housing assembly".

**There is lack of antecedent basis for "the surface" and "the end" recited in claim 8.** Claim 8 has been canceled. All its texts including "the surface" and "the end" no longer exist in the amended claims.

**It is unclear which cylindrical ring is being referred to in claim 9.** Claim 9 has been canceled. All its texts no longer exist in the amended claims.

**It is unclear what is meant by the "embed location is in-between a cut off portion of said cylindrical ring and said cylindrical ring itself" in claim 9.** Claim 9 has been canceled. All its texts no longer exist in the amended claims.

**The language "prepares itself for future changes and improvements" in claim 10 is unclear and indefinite.** Claim 10 is now claim 14 in the amended claims. Claim 14 uses the language "allows said digital camera to accept new picture-taking capacities that are introduced in the form of adapter rings" to replace the language "prepares itself for future changes and improvements".

**It is unclear what is mean by "not normally common" as recited in claim 10.** Claim 10 is now claim 14 in the amended claims. Claim 14 uses the language "to said digital camera beside that of point-and-shoot method" to replace the language "that are not normally common with a digital camera".

**There is lack of antecedent basis for "the eyepiece" recited in claim 10.** Claim 10 is now claim 14 in the amended claims. In the "whereby" paragraph of claim 14 "the eyepiece" has been changed to "said eyepiece" to refer to an eyepiece of a sighting device in the main paragraph of claim 14.

**Response 15**

**Response 16**

**Response 17**

## Claim Rejections – 35 USC § 102

**Claims 1-3, 5-7 and 9-10 are rejected under 35 U.S.C. 102(a)(1) as being anticipated by Nelson et al. (U.S. Patent No. 3,752,569).**
Claims 1-10 have all been canceled and replaced with claims 11-14.

The applicant's proposed invention targets a small digital camera measuring approximately 3.5" wide x 1" high x 2.5" deep and weighing approximately 2.5 ounces. When mounted on the digital camera, the invention adds a unique and innovative enhancement to the camera's picture- and motion-taking capacities.

The camera mount assembly having a circular base and a cylindrical ring 90 referred to in Nelson et al is a complex invention designed for a motion picture projector using films. The motion picture projector is significantly larger than a digital camera. Nelson et al's invention is impossible to mount to a digital camera of such a small size. The economic justification for doing so does not exist. The digital camera has already a motion picture-taking capacity in the first place. The camera mount assembly, being unworkable with a digital camera, renders the accompanying main housing assembly unworkable as well.

According to Wikipedia, the first digital camera of any kind ever sold commercially was possibly the MegaVision Tessera in 1987, 15 years later than the year (1973) the patent for the camera mount assembly was issued to Nelson et al. In 1973 the form factor and size in which the digital camera would come into being was unknown. In the 15-year gap applicant has not seen a digital camera of the aforementioned size equipped with Nelson et al's camera mount assembly.

Applicant respectfully argues that there was no novelty introduced by the camera mount assembly referred to in Nelson et al to the digital camera that is the object of the applicant's proposed invention. On the other hand, applicant's proposed invention introduces more picture-taking capacities to the digital camera.

**The language "for providing enclosure and track"** which appears in claim 1 (now claim 11 in the amended claims) subparagraph a, has now been changed to "for allowing unobstructed movement of said digital camera's barrel".

**The language "for accepting and securing"** which appears in claim 1 (now claim 11 in the amended claims) subparagraph b, has now been changed to "for union with camera filters and sighting devices".

**Response 18**

**Response 19**

**Response 20**

The circular base and the cylindrical ring have a locking method for engaging said circular base and said cylindrical ring together as shown in figure 4.

The digital camera is the object of the applicant's proposed invention. Applicant respectfully invokes the same arguments in the previous paragraphs referring to the unworkable nature of Nelson et al's invention towards the digital camera.

**Regarding claim 6, the cylindrical ring of the main housing assembly has more than one embed location. One embed location is in the front (right side in figure 4) and one embed location is in the back (left side in figure 4).**

Claim 6 has been canceled. All its texts including reference to embed locations no longer exist in the amended claims.

### Claim Rejections – 35 USC § 103

**Claim 4 is rejected under 35 U.S.C. 103 as being unpatentable over Nelson et al. (U.S. Patent No. 3,752,569) in view of Spencer (.U.S. Paent No. 79088918).**

Claim 4 no longer exists in the amended claims. However, the following argument is still valid and helpful in seeking a grant of a patent to applicant's proposed invention.

In previous paragraphs the applicant has argued that the camera mount assembly having a circular base and a cylindrical ring 90 referred to in Nelson et al is not workable with the digital camera that is the object of the applicant's proposed invention.

The threaded connector for lens adapters referred to in view of Spencer (U.S. Patent No. 7088918) is specifically designed to work with a camera with automatic exposure mode. The following quote from the SUMMARY OF THE INVENTION partly explains how the threaded connector is made to work with the camera. "The lens of the camera 2 with automatic exposure mode **is removed** in order to amount the camera 2 with the adapter 20 to the optical system 4"

It is apparent that the camera with automatic exposure mode is of the type which has a removable lens. It is impossible even for one of ordinary skill in the art to use the features of Spencer for the purpose of quick connection on a digital camera referred to in the applicant's proposed invention. The reason is that the digital camera **has a fixed lens** rendering adapter 20 unusable.

Therefore, applicant requests that the threaded connector for lens adapters referred to in view of Spencer (U.S. Patent No. 7088918) be not considered prior art with respect to the digital camera that is the object of the applicant's proposed invention.

**Response 21**

**Claim 8 is rejected under 35 U.S.C. 103 as being unpatentable over Nelson et al. (U.S. Patent No. 3,752,569) in view of MacKay (U.s. Patent No. 5208624).**
Claim 8 no longer exists in the amended claims. However, the following argument is still valid and helpful in seeking a grant of a patent to applicant's proposed invention.

In previous paragraphs the applicant has argued that the camera mount assembly having a circular base and a cylindrical ring 90 referred to in Nelson et al is not workable with the digital camera that is the object of the applicant's proposed invention.

MacKay teaches that it was known to embed the adapter on the surface via magnets. On a digital camera measuring approximately 3.5" wide x 1" high x 2.5", which is the object of the applicant's proposed invention, a strong magnet will be required to mount the adapter to the camera.

A strong magnet has been known to interfere with a digital camera's electronics and storage. It also interferes with someone's pacemaker if a digital camera attached with a strong magnet is inserted in a shirt pocket - a habit frequently and unconsciously done.

For the above reasons, one of ordinary skill in the art would not make the features of MacKay for the purpose of quick disconnects, work with a digital camera that is the object of the applicant's proposed invention.

Therefore, applicant respectfully argues that the features of MacKay be not considered prior art with respect to a digital camera and requests that the above basis for rejection be withdrawn.

Response - Amendment A - Conclusion

Appn. Number 14/028,786    (Roy Lique)    Art Unit 2852    Amnt. A    Page 10 of 10

**CONCLUSION**

For all the above reasons, applicant submits that the claims are now in proper form and that the claims all define patentability over the prior art. Therefore he submits that this application is now in condition for allowance, which action he respectfully solicits.

**Conditional Request for Constructive Assistance**

Applicant has amended the claims of this application so that they are proper, definite, and define novel structure which is also unobvious. If, for any reason this application is not believed to be in full condition for allowance, applicant respectfully requests the constructive assistance and suggestions of the Examiner pursuant to M.P.E.P § 2173.02 and § 707.07(j) in order that the undersigned can place this application in allowable condition as soon as possible and without the need for further proceedings.

Very respectfully,

Roy Lique

1431 Tierra Cima
Walnut, CA 91789
Tel. (909) 594-1470, Fax (909) 594-1470

Certificate of Facsimile Transmission. I certify that on the date below I will fax this paper to Art Unit2852 of the U.S. Patent and Trademarks Office at 571-273-8300.

July 30, 2015          _____

# Final Office Action

## Office Action Summary

**3 months to reply**

**Response to previous communication... Office action is final**

**Claims 11-14 are pending. Claims 11, 13, and 14 allowed. Claim 12 is rejected.**

---

| | | Application No. 14/028.786 | Applicant(s) LIQUE, ROY | |
|---|---|---|---|---|
| **Office Action Summary** | | Examiner Christopher Mahoney | Art Unit 2852 | AIA (First Inventor to File) Status Yes |

-- The MAILING DATE of this communication appears on the cover sheet with the correspondence address --

**Period for Reply**

A SHORTENED STATUTORY PERIOD FOR REPLY IS SET TO EXPIRE _3_ MONTHS FROM THE MAILING DATE OF THIS COMMUNICATION.

- Extensions of time may be available under the provisions of 37 CFR 1.136(a). In no event, however, may a reply be timely filed after SIX (6) MONTHS from the mailing date of this communication.
- If NO period for reply is specified above, the maximum statutory period will apply and will expire SIX (6) MONTHS from the mailing date of this communication.
- Failure to reply within the set or extended period for reply will, by statute, cause the application to become ABANDONED (35 U.S.C. § 133).
- Any reply received by the Office later than three months after the mailing date of this communication, even if timely filed, may reduce any earned patent term adjustment. See 37 CFR 1.704(b).

**Status**

1) ☒ Responsive to communication(s) filed on _July 30, 2015_.
   ☐ A declaration(s)/affidavit(s) under **37 CFR 1.130(b)** was/were filed on _____.
2a) ☒ This action is **FINAL**.    2b) ☐ This action is non-final.
3) ☐ An election was made by the applicant in response to a restriction requirement set forth during the interview on _____; the restriction requirement and election have been incorporated into this action.
4) ☐ Since this application is in condition for allowance except for formal matters, prosecution as to the merits is closed in accordance with the practice under *Ex parte Quayle*, 1935 C.D. 11, 453 O.G. 213.

**Disposition of Claims***

5) ☒ Claim(s) _11-14_ is/are pending in the application.
   5a) Of the above claim(s) _____ is/are withdrawn from consideration.
6) ☒ Claim(s) _11,13 and 14_ is/are allowed.
7) ☒ Claim(s) _12_ is/are rejected.
8) ☐ Claim(s) _____ is/are objected to.
9) ☐ Claim(s) _____ are subject to restriction and/or election requirement.

* If any claims have been determined _allowable_, you may be eligible to benefit from the **Patent Prosecution Highway** program at a participating intellectual property office for the corresponding application. For more information, please see http://www.uspto.gov/patents/init_events/pph/index.jsp or send an inquiry to PPHfeedback@uspto.gov.

**Application Papers**

10) ☐ The specification is objected to by the Examiner.
11) ☐ The drawing(s) filed on _____ is/are: a) ☐ accepted or b) ☐ objected to by the Examiner.
   Applicant may not request that any objection to the drawing(s) be held in abeyance. See 37 CFR 1.85(a).
   Replacement drawing sheet(s) including the correction is required if the drawing(s) is objected to. See 37 CFR 1.121(d).

**Priority under 35 U.S.C. § 119**

12) ☐ Acknowledgment is made of a claim for foreign priority under 35 U.S.C. § 119(a)-(d) or (f).
   **Certified copies:**
   a) ☐ All    b) ☐ Some**    c) ☐ None of the:
   1. ☐ Certified copies of the priority documents have been received.
   2. ☐ Certified copies of the priority documents have been received in Application No. _____.
   3. ☐ Copies of the certified copies of the priority documents have been received in this National Stage application from the International Bureau (PCT Rule 17.2(a)).
   ** See the attached detailed Office action for a list of the certified copies not received.

**Attachment(s)**

1) ☒ Notice of References Cited (PTO-892)
2) ☐ Information Disclosure Statement(s) (PTO/SB/08a and/or PTO/SB/08b) Paper No(s)/Mail Date _____.
3) ☐ Interview Summary (PTO-413) Paper No(s)/Mail Date _____.
4) ☐ Other _____.

U.S. Patent and Trademark Office
PTOL-326 (Rev. 11-12)        Office Action Summary        Part of Paper No./Mail Date 51

## Detailed Action – Page 1

**Detailed Action 1**

Application/Control Number: 14/028,786                                   Page 2
Art Unit: 2852

The present application, filed on or after March 16, 2013, is being examined under the

first inventor to file provisions of the AIA.

### DETAILED ACTION

#### *Claim Rejections - 35 USC § 112*

The following is a quotation of 35 U.S.C. 112(b):

(b) CONCLUSION.   The specification shall conclude with one or more claims particularly pointing
out and distinctly claiming the subject matter which the inventor or a joint inventor regards as the
invention.

The following is a quotation of 35 U.S.C. 112 (pre-AIA), second paragraph:

The specification shall conclude with one or more claims particularly pointing and distinctly
claiming the subject matter which the applicant regards as his invention.

The following is a quotation of 35 U.S.C. 112(d):

(d) REFERENCE IN DEPENDENT FORMS.— Subject to subsection (e), a claim in dependent form
shall contain a reference to a claim previously set forth and then specify a further limitation of the
subject matter claimed. A claim in dependent form shall be construed to incorporate by reference all the
limitations of the claim to which it refers.

The following is a quotation of 35 U.S.C. 112 (pre-AIA), fourth paragraph:

Subject to the [fifth paragraph of 35 U.S.C. 112 (pre-AIA)], a claim in dependent form shall contain a
reference to a claim previously set forth and then specify a further limitation of the subject matter
claimed. A claim in dependent form shall be construed to incorporate by reference all the limitations of
the claim to which it refers.

Claim 12 is rejected under 35 U.S.C. 112(b) or 35 U.S.C. 112 (pre-AIA), second

paragraph, as being indefinite for failing to particularly point out and distinctly claim the subject

matter which the inventor or a joint inventor, or for pre-AIA the applicant regards as the

invention.

Claim 12 recites that the camera mount assembly being fitted with an adapter ring, for

directly adding picture taking capacities to said digital camera without using said primary ring of

# Detailed Action – Page 2

said main housing assembly. It is unclear if the main housing, the cylindrical ring or both are being removed by claim 12. In any case, this renders the claim indefinite since claim 11 from which claim 12 depends, positively recites those elements.

## *Allowable Subject Matter*

**Detailed Action 2**

**Claims 11, 13, and 14 are allowed.**

Claims 11 and 13-14 are allowed.

The prior art teaches a digital camera lens guard and use extender comprising a camera mount assembly having a circular base and a cylindrical ring of sufficient size and length and a plurality of locking means, for allowing unobstructed movement of said digital cameras barrel. See for example, Sakurai (U.S. Publication No. 2008/0205881) and Yang (U.S. Publication No. 2012/0236424). Sakurai teaches a plurality of locking means 52/44/46, fig 7A. Both Sakurai and Yang teach a main housing assembly. The prior art does not teach, in combination with the additionally recited elements, a main housing assembly a main housing assembly having a primary ring of sufficient size and length with a plurality of retaining screws and embed locations for adapter rings on each end, for union with camera filters and sighting devices.

**Claim 12 allowable if rewritten to overcome rejection**

Claim 12 would be allowable if rewritten to overcome the rejection(s) under 35 U.S.C. 112(b) or 35 U.S.C. 112 (pre-AIA), 2nd paragraph, set forth in this Office action and to include all of the limitations of the base claim and any intervening claims.

Nelson does not teach attachment to a camera. Further, the lenses in Nelson are meant to be stationary.

## *Response to Arguments*

# Detailed Action – Page 3

Application/Control Number: 14/028,786                                          Page 4
Art Unit: 2852

Applicant's arguments with respect to claim 12 have been considered but are moot because the arguments do not apply to any of the references being used in the current rejection.

The applicant's arguments on page 8 of the remarks regarding the time difference of the Nelson reference and modern digital cameras are not persuasive. One of ordinary skill in the art could take the teachings of prior art and apply them to new technology. However Nelson is silent as to having a sufficient size and length to allow unobstructed movement of a camera barrel. The optics of Nelson are designed to be more stationary.

*Conclusion*

Any inquiry concerning this communication or earlier communications from the examiner should be directed to Christopher Mahoney whose telephone number is (571)272-2122. The examiner can normally be reached on 8:30AM-5PM, Monday-Thursday.

If attempts to reach the examiner by telephone are unsuccessful, the examiner's supervisor, Clayton Laballe can be reached on (571) 272-1594. The fax phone number for the organization where this application or proceeding is assigned is 571-273-8300.

**Detailed Action – Page 4**

Application/Control Number: 14/028,786                                    Page 5
Art Unit: 2852

Information regarding the status of an application may be obtained from the Patent

Application Information Retrieval (PAIR) system. Status information for published applications

may be obtained from either Private PAIR or Public PAIR. Status information for unpublished

applications is available through Private PAIR only. For more information about the PAIR

system, see http://pair-direct.uspto.gov. Should you have questions on access to the Private PAIR

system, contact the Electronic Business Center (EBC) at 866-217-9197 (toll-free). If you would

like assistance from a USPTO Customer Service Representative or access to the automated

information system, call 800-786-9199 (IN USA OR CANADA) or 571-272-1000.


/Christopher Mahoney/
Primary Examiner, Art Unit 2852

**Interference Search**

| INTERFERENCE SEARCH | | | |
|---|---|---|---|
| US Class/ CPC Symbol | US Subclass / CPC Group | Date | Examiner |
| G03B | 17/12 | 10/30/2015 | CM |
| G03B | 17/14 | 10/30/2015 | CM |
| G03B | 17/565 | 10/30/2015 | CM |
| H04N | 5/2254 | 10/30/2015 | CM |
| 396 | 71 | 10/30/2015 | CM |
| 396 | 544 | 10/30/2015 | CM |
| 396 | 611 | 10/30/2015 | CM |
| 396 | 612 | 10/30/2015 | CM |
| 359 | 811 | 10/30/2015 | CM |
| 359 | 819 | 10/30/2015 | CM |
| 359 | 822 | 10/30/2015 | CM |
| 359 | 826 | 10/30/2015 | CM |
| 359 | 823 | 10/30/2015 | CM |

U.S. Patent and Trademark Office

Part of Paper No. 51

# Notice of References Cited

| | | Application/Control No. | Applicant(s)/Patent Under Reexamination | | |
|---|---|---|---|---|---|
| **Notice of References Cited** | | 14/028,786 | LIQUE, ROY | | |
| | | Examiner | Art Unit | | |
| | | Christopher Mahoney | 2852 | Page 1 of 1 | |

**U.S. PATENT DOCUMENTS**

| * | | Document Number Country Code-Number-Kind Code | Date MM-YYYY | Name | CPC Classification | US Classification |
|---|---|---|---|---|---|---|
| * | A | US-2012/0236424 A1 | 09-2012 | Yang; Chih-Yi | G02B7/022 | 359/819 |
| * | B | US-2008/0205881 A1 | 08-2008 | Sakurai; Nobumasa | G02B7/14 | 396/530 |
| | C | US- | | | | |
| | D | US- | | | | |
| | E | US- | | | | |
| | F | US- | | | | |
| | G | US- | | | | |
| | H | US- | | | | |
| | I | US- | | | | |
| | J | US- | | | | |
| | K | US- | | | | |
| | L | US- | | | | |
| | M | US- | | | | |

**FOREIGN PATENT DOCUMENTS**

| * | | Document Number Country Code-Number-Kind Code | Date MM-YYYY | Country | Name | CPC Classification |
|---|---|---|---|---|---|---|
| | N | | | | | |
| | O | | | | | |
| | P | | | | | |
| | Q | | | | | |
| | R | | | | | |
| | S | | | | | |
| | T | | | | | |

**NON-PATENT DOCUMENTS**

| * | | Include as applicable: Author, Title Date, Publisher, Edition or Volume, Pertinent Pages) |
|---|---|---|
| | U | |
| | V | |
| | W | |
| | X | |

*A copy of this reference is not being furnished with this Office action. (See MPEP § 707.05(a).)
Dates in MM-YYYY format are publication dates. Classifications may be US or foreign.

U.S. Patent and Trademark Office
PTO-892 (Rev. 01-2001) — Notice of References Cited — Part of Paper No. 51

## Search Notes

| Search Notes | Application/Control No. | Applicant(s)/Patent Under Reexamination |
|---|---|---|
| | 14028786 | LIQUE, ROY |
| | Examiner | Art Unit |
| | CHRIST MAHONEY | 2852 |

### CPC- SEARCHED

| Symbol | Date | Examiner |
|---|---|---|
| G03B 17/12 | 4/28/2015 | CM |
| G03B 17/14 | 4/28/2015 | CM |
| G03B 17/565 | 4/28/2015 | CM |
| H04N 5/2254 | 4/28/2015 | CM |
| updated search | 10/30/2015 | CM |

### CPC COMBINATION SETS - SEARCHED

| Symbol | Date | Examiner |
|---|---|---|
| | | |

### US CLASSIFICATION SEARCHED

| Class | Subclass | Date | Examiner |
|---|---|---|---|
| 396 | 71 | 4/28/2015 | CM |
| 396 | 544 | 4/28/2015 | CM |
| 396 | 611 | 4/28/2015 | CM |
| 396 | 612 | 4/28/2015 | CM |
| 359 | 811 | 4/28/2015 | CM |
| 359 | 819 | 4/28/2015 | CM |
| 359 | 822 | 4/28/2015 | CM |
| 359 | 826 | 4/28/2015 | CM |
| 359 | 823 | 4/28/2015 | CM |
| updated | search | 10/30/2015 | CM |

### SEARCH NOTES

| Search Notes | Date | Examiner |
|---|---|---|
| PLUS search | 4/03/2015 | CM |
| EAST | 4/28/2015 | CM |
| EAST; inventor name search | 10/30/2015 | CM |

U.S. Patent and Trademark Office

Part of Paper No. : 51

# Response - Amendment B

## Amendment B - Cover Letter

### In the United States Patent and Trademark Office

| | |
|---|---|
| Appn. Number: | 14/028,786 |
| Appn. Filed | 09/17/2013 |
| Applicants: | Roy Lique |
| Title: | Digital Camera Lens Guard and Use Extender |
| Examiner/GAU: | Christopher Mahoney/Art Unit 2852 |

Chino Hills, January 6, 2016

### AMENDMENT B

Mail Stop Non-Fee Amendments
Commissioner for Patents
P.O. Box 1450
Alexandria, VA 22313-1450

Sir:

In response to the Office Action Mailed November 4, 2015, please amend the above application as follows:

CLAIMS: Amendments to the claims begin on page 2 of this Amendment
REMARKS: Remarks begin on page 3 of this Amendment.

**CLAIMS**

Please amend the claims according to the status designations in the following list, which contains changes of all claims that were ever in the application, with the text of all new active claims.

CLAIMS: I claim:

11. (PREVIOUSLY PRESENTED) A digital camera lens guard and use extender for adding deviant picture-taking capacities to a digital camera, and protecting said camera's lens, barrel, and electronics, comprising:

    a. a camera mount assembly having a circular base and a cylindrical ring of sufficient size and length and a plurality of locking means, for allowing unobstructed movement of said digital camera's barrel,

    b. a main housing assembly having a primary ring of sufficient size and length with a plurality of retaining screws and embed locations for adapter rings on each end, for union with camera filters and sighting devices.

12. (CANCELED)

13. (PREVIOUSLY PRESENTED) The digital camera lens guard and use extender of claim 11 wherein said embed location is a circular recess carved approximately one-eighth to one-fourth inch at each said end of said primary ring of said main housing assembly.

14. (PREVIOUSLY PRESENTED) The digital camera lens guard and use extender of claim 11 wherein said primary ring of said main housing assembly being inserted with a coupler ring, for securing an eyepiece of a sighting device,

whereby the digital camera lens guard and use extender adds picture-taking capacities to said digital camera beside that of point-and-shoot method; provides protection for said digital camera's lens, barrel, and electronics; facilitates usage of camera filters, adapter rings, and sighting devices; enables said digital camera to take pictures of images formed at said eyepiece of a sighting device; and allows said digital camera to accept new picture-taking capacities that are introduced in the form of adapter rings.

## CLAIMS

Please amend the claims according to the status designations in the following list, which contains changes of all claims that were ever in the application, with the text of all new active claims.

CLAIMS: I claim:

11. (PREVIOUSLY PRESENTED) A digital camera lens guard and use extender for adding deviant picture-taking capacities to a digital camera, and protecting said camera's lens, barrel, and electronics, comprising:

    a. a camera mount assembly having a circular base and a cylindrical ring of sufficient size and length and a plurality of locking means, for allowing unobstructed movement of said digital camera's barrel,

    b. a main housing assembly having a primary ring of sufficient size and length with a plurality of retaining screws and embed locations for adapter rings on each end, for union with camera filters and sighting devices.

12. (CANCELED)

13. (PREVIOUSLY PRESENTED) The digital camera lens guard and use extender of claim 11 wherein said embed location is a circular recess carved approximately one-eighth to one-fourth inch at each said end of said primary ring of said main housing assembly.

14. (PREVIOUSLY PRESENTED) The digital camera lens guard and use extender of claim 11 wherein said primary ring of said main housing assembly being inserted with a coupler ring, for securing an eyepiece of a sighting device,

whereby the digital camera lens guard and use extender adds picture-taking capacities to said digital camera  beside that of point-and-shoot method; provides protection for said digital camera's lens, barrel, and electronics; facilitates usage of camera filters, adapter rings, and sighting devices; enables said digital camera to take pictures of images formed at said eyepiece of a sighting device; and allows said digital camera to accept new picture-taking capacities that are introduced in the form of adapter rings.

## Response to Detailed Action

1.

I preferred not to rewrite claim 12 as offered.

Appn. Number 14/028,786    (Roy Lique)    Art Unit 2852    Amnt. B    Page 3 of 4

**REMARKS – General**

By the above amendment, applicant has corrected, to the best of his knowledge, the following:

**Claim 12 is rejected under 35 US.C. 112(b or 35 U.S.C. 112 (pre-AIA), second paragraph, as being indefinite for failing to particularly point out and distinctly claim the subject matter which the inventor or a joint inventor, for pre-AIA the applicant regards as the invention.**

Claim 12 has been canceled. All its texts no longer exist in the amended claims.

Appn. Number 14/028,786     (Roy Lique)     Art Unit 2852     Amnt. B     Page **4** of **4**

## CONCLUSION

For all the above reasons, applicant submits that the claims are now in proper form and that the claims all define patentability over the prior art. Therefore he submits that this application is now in condition for allowance, which action he respectfully solicits.

### Conditional Request for Constructive Assistance

Applicant has amended the claims of this application so that they are proper, definite, and define novel structure which is also unobvious. If, for any reason this application is not believed to be in full condition for allowance, applicant respectfully requests the constructive assistance and suggestions of the Examiner pursuant to M.P.E.P § 2173.02 and § 707.07(j) in order that the undersigned can place this application in allowable condition as soon as possible and without the need for further proceedings.

Very respectfully,

Roy Lique

2650 Lookout Cir.
Chino Hills, CA 91709
Tel. (909) 591-2529, Fax (909) 591-2529

Certificate of Facsimile Transmission. I certify that on the date below I will fax this paper to Art Unit 2852 of the U.S. Patent and Trademarks Office at 571-273-8300.

January 6, 2016     _____

# Part 7 – Grant of Patent

## Notice of Allowability – Provisions of the law

### Allowance and Issue of Patent
If, on examination of the application, or at a later stage during the reconsideration of the application, the patent application is found to be allowable, a Notice of Allowance and Fee(s) Due will be sent to the applicant, or to applicant's attorney or agent of record, if any, and a fee for issuing the patent and if applicable, for publishing the patent application publication (see 37 CFR 1.211-1.221), is due within three months from the date of the notice. If timely payment of the fee(s) is not made, the application will be regarded as abandoned. See the current fee schedule at www.uspto.gov. The Director may accept the fee(s) late, if the delay is shown to be unavoidable (35 U.S.C. 41, 37 CFR 1.137(a)) or unintentional (35 U.S.C. 151, 37 CFR 1.137(b)). When the required fees are paid, the patent issues as soon as possible after the date of payment, dependent upon the volume of printing on hand. The patent grant then is delivered or mailed on the day of its grant, or as soon thereafter as possible, to the inventor's attorney or agent if there is one of record, otherwise directly to the inventor. On the date of the grant, the patent file becomes open to the public for applications not opened earlier by publication of the application.

In cases where the publication of an application or the granting of a patent would be detrimental to the national security, the Commissioner for Patents will order that the invention be kept secret and shall withhold the publication of the application or the grant of the patent for such period as the national interest requires. The owner of an application that has been placed under a secrecy order has a right to appeal the order to the Secretary of Commerce. 35 U.S.C. 181.

### Patent Term Extension and Adjustment
The terms of certain patents may be subject to extension or adjustment under 35 U.S.C. 154(b). Such extension or adjustment results from certain specified types of delays which may occur while an application is pending before the Office.

Utility and plant patents which issue from original applications filed between June 8, 1995 and May 28, 2000 may be eligible for patent term extension (PTE) as set forth in 37 CFR 1.701. Such PTE may result from delays due to interference proceedings under 35 U.S.C. 135(a), secrecy orders under 35 U.S.C. 181, or successful appellate review.

Utility and plant patents which issue from original applications filed on or after May 29, 2000 may be eligible for patent term adjustment (PTA) as set forth in 37 CFR 1.702 - 1.705. There are three main bases for PTA under 35 U.S.C. 154(b). The first basis for PTA is the failure of the Office to take certain actions within specific time frames set forth in 35 U.S.C. 154(b)(1)(A) (See 37 CFR 1.702(a) and 1.703(a)). The second basis for PTA is the failure of the Office to issue a patent within three years of the actual filing date of the application as set forth in 35 U.S.C. 154(b)(1)(B) (See 37 CFR 1.702(b) and 1.703(b)). The third basis for PTA is set forth in 35 U.S.C. 154(b)(1)(C), and includes delays due to interference proceedings under 35 U.S.C. 135(a), secrecy orders under 35 U.S.C. 181, or successful appellate review (See 37 CFR 1.702(c)-(e) and 1.703(c)-(e)).

Any PTA which has accrued in an application will be reduced by the time period during which an applicant failed to engage in reasonable efforts to conclude prosecution of the application pursuant to 35 U.S.C. 154(b)(2)(C). A non-exclusive list of activities which constitute failure to engage in reasonable efforts to conclude prosecution is set forth in 37 CFR 1.704.24

An initial PTA value is printed on the notice of allowance and fee(s) due, and a final PTA value is printed on the front of the patent. Any request for reconsideration of the PTA value printed on the notice of allowance and fee(s) due should be made in the form of an application for patent term adjustment, which must be filed prior to or at the same time as the payment of the issue fee. (See 37 CFR 1.705.)

## Nature of Patent and Patent Rights

The patent is issued in the name of the United States under the seal of the United States Patent and Trademark Office, and is either signed by the Director of the USPTO or is electronically written thereon and attested by an Office official. The patent contains a grant to the patentee, and a printed copy of the specification and drawing is annexed to the patent and forms a part of it. The grant confers "the right to exclude others from making, using, offering for sale, or selling the invention throughout the United States or importing the invention into the United States" and its territories and possessions for which the term of the patent shall be generally 20 years from the date on which the application for the patent was filed in the United States or, if the application contains a specific reference to an earlier filed application under 35 U.S.C. 120, 121 or 365(c), from the date of the earliest such application was filed, and subject to the payment of maintenance fees as provided by law.

The exact nature of the right conferred must be carefully distinguished, and the key is in the words "right to exclude" in the phrase just quoted. The patent does not grant the right to make, use, offer for sale or sell or import the invention but only grants the exclusive nature of the right. Any person is ordinarily free to make, use, offer for sale or sell or import anything he or she pleases, and a grant from the government is not necessary. The patent only grants the right to exclude others from making, using, offering for sale or selling or importing the invention. Since the patent does not grant the right to make, use, offer for sale, or sell, or import the invention, the patentee's own right to do so is dependent upon the rights of others and whatever general laws might be applicable. A patentee, merely because he or she has received a patent for an invention, is not thereby authorized to make, use, offer for sale, or sell, or import the invention if doing so would violate any law.

An inventor of a new automobile who has obtained a patent thereon would not be entitled to use the patented automobile in violation of the laws of a state requiring a license, nor may a patentee sell an article, the sale of which may be forbidden by a law, merely because a patent has been obtained.

Neither may a patentee make, use, offer for sale, or sell, or import his or her own invention if doing so would infringe the prior rights of others. A patentee may not violate the federal antitrust laws, such as by resale price agreements or entering into combination in restraints of trade, or the pure food and drug laws, by virtue of having a patent. Ordinarily there is nothing that prohibits a patentee from making, using, offering for sale, or selling, or importing his or her own invention, unless he or she thereby infringes another's patent that is still in force. For example, a patent for an improvement of an original device already patented would be subject to the patent on the device.

The term of the patent shall be generally 20 years from the date on which the application for the patent was filed in the United States or, if the application contains a specific reference to an earlier filed application under 35 U.S.C. 120, 121 or 365(c), from the date of the earliest such application was filed, and subject to the payment of maintenance fees as provided by law. A maintenance fee is due 3.5, 7.5 and 11.5 years after the original grant for all patents issuing from the applications filed on and after December 12, 1980. The maintenance fee must be paid at the stipulated times to maintain the patent in force. After the patent has expired anyone may make, use, offer for sale, or sell or import the invention without permission of the patentee, provided that matter covered by other unexpired patents is not used. The terms may be extended for certain pharmaceuticals and for certain circumstances as provided by law.

## Maintenance Fees

All utility patents that issue from applications filed on or after December 12, 1980 are subject to the payment of maintenance fees which must be paid to maintain the patent in force. These fees are due at 3.5, 7.5 and 11.5 years from the date the patent is granted and can be paid without a surcharge during the "window period," which is the six-month period preceding each due date, e.g., three years to three years and six months. (See fee schedule for a list of maintenance fees.) In submitting maintenance fees and any necessary surcharges, identification of the patents for which maintenance fees are being paid must include the patent number, and the application number of the U.S. application for the patent on which the maintenance fee is being paid. If the payment includes identification of only the patent number, the Office may apply payment to the patent identified by patent number in the payment or the Office may return the payment. (See 37, Code of Federal Regulations, section 1.366(c).)

Failure to pay the current maintenance fee on time may result in expiration of the patent. A six-month grace period is provided when the maintenance fee may be paid with a surcharge. The grace period is the six-month period immediately following the due date. The USPTO does not mail notices to patent owners that maintenance fees are due. If, however, the maintenance fee is not paid on time, efforts are made to remind the responsible party that the maintenance fee may be paid during the grace period with a surcharge. If the maintenance fee is not paid on time and the maintenance fee and surcharge are not paid during the grace period, the patent expires on the date the grace period ends.

## Notice of Allowability

Claims 11, 13, and 14 are allowed.

Claim 12 was cancelled.

Claims are renumbered in final issuance of the patent certificate.

| | |
|---|---|
| **Notice of Allowability** | **Application No.** 14/028,786 / **Applicant(s)** LIQUE, ROY |
| | **Examiner** Christopher Mahoney / **Art Unit** 2852 / **AIA (First Inventor to File) Status** Yes |

-- *The MAILING DATE of this communication appears on the cover sheet with the correspondence address--*
All claims being allowable, PROSECUTION ON THE MERITS IS (OR REMAINS) CLOSED in this application. If not included herewith (or previously mailed), a Notice of Allowance (PTOL-85) or other appropriate communication will be mailed in due course. **THIS NOTICE OF ALLOWABILITY IS NOT A GRANT OF PATENT RIGHTS.** This application is subject to withdrawal from issue at the initiative of the Office or upon petition by the applicant. See 37 CFR 1.313 and MPEP 1308.

1. ☒ This communication is responsive to *the amendment filed January 11, 2016.*
    ☐ A declaration(s)/affidavit(s) under **37 CFR 1.130(b)** was/were filed on ____.

2. ☐ An election was made by the applicant in response to a restriction requirement set forth during the interview on ____; the restriction requirement and election have been incorporated into this action.

3. ☒ The allowed claim(s) is/are *11,13 and 14*. As a result of the allowed claim(s), you may be eligible to benefit from the **Patent Prosecution Highway** program at a participating intellectual property office for the corresponding application. For more information, please see http://www.uspto.gov/patents/init_events/pph/index.jsp or send an inquiry to PPHfeedback@uspto.gov.

4. ☐ Acknowledgment is made of a claim for foreign priority under 35 U.S.C. § 119(a)-(d) or (f).
    **Certified copies:**
    a) ☐ All    b) ☐ Some   *c) ☐ None of the:
        1. ☐ Certified copies of the priority documents have been received.
        2. ☐ Certified copies of the priority documents have been received in Application No. ____.
        3. ☐ Copies of the certified copies of the priority documents have been received in this national stage application from the International Bureau (PCT Rule 17.2(a)).
    * Certified copies not received: ____.

Applicant has THREE MONTHS FROM THE "MAILING DATE" of this communication to file a reply complying with the requirements noted below. Failure to timely comply will result in ABANDONMENT of this application.
THIS THREE-MONTH PERIOD IS NOT EXTENDABLE.

5. ☐ CORRECTED DRAWINGS ( as "replacement sheets") must be submitted.
    ☐ including changes required by the attached Examiner's Amendment / Comment or in the Office action of Paper No./Mail Date ____.
    **Identifying indicia** such as the application number (see 37 CFR 1.84(c)) should be written on the drawings in the front (not the back) of each sheet. Replacement sheet(s) should be labeled as such in the header according to 37 CFR 1.121(d).

6. ☐ DEPOSIT OF and/or INFORMATION about the deposit of BIOLOGICAL MATERIAL must be submitted. Note the attached Examiner's comment regarding REQUIREMENT FOR THE DEPOSIT OF BIOLOGICAL MATERIAL.

Attachment(s)
1. ☐ Notice of References Cited (PTO-892)
2. ☐ Information Disclosure Statements (PTO/SB/08), Paper No./Mail Date ____
3. ☐ Examiner's Comment Regarding Requirement for Deposit of Biological Material
4. ☐ Interview Summary (PTO-413), Paper No./Mail Date ____

5. ☐ Examiner's Amendment/Comment
6. ☐ Examiner's Statement of Reasons for Allowance
7. ☐ Other ____

/Christopher Mahoney/
Primary Examiner, Art Unit 2852

U.S. Patent and Trademark Office
PTOL-37 (Rev. 08-13)               Notice of Allowability          Part of Paper No./Mail Date 62

**Amendment B Cover Letter attached by USPTO**

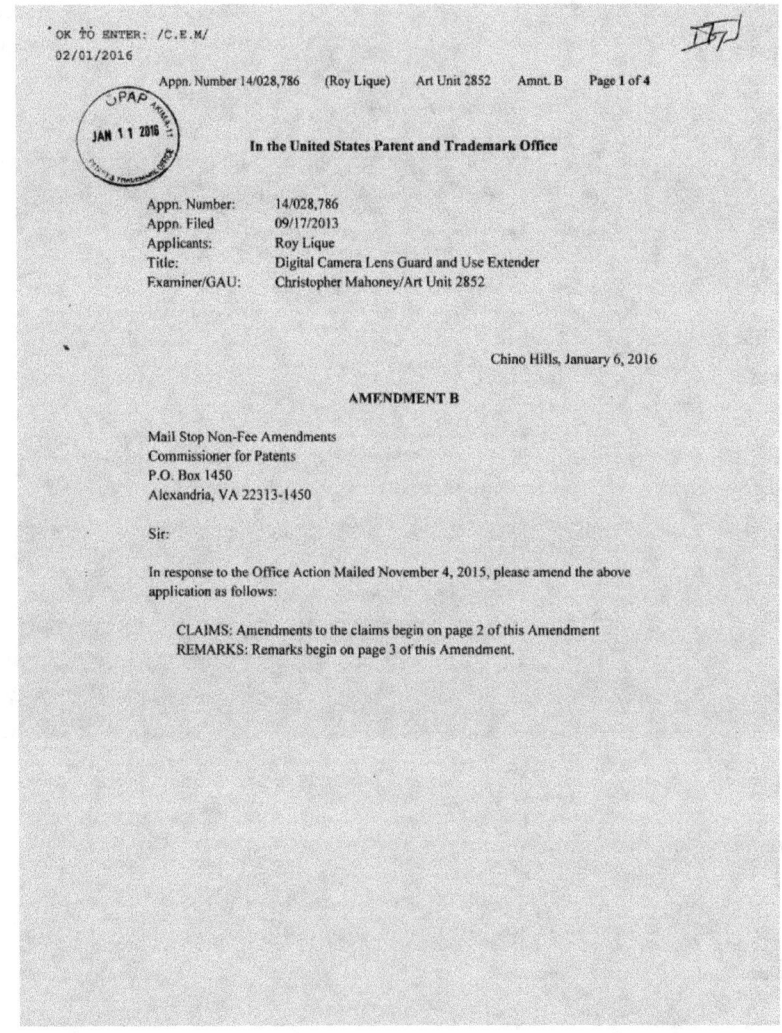

OK TO ENTER: /C.E.M/
02/01/2016

Appn. Number 14/028,786    (Roy Lique)    Art Unit 2852    Amnt. B    Page 1 of 4

JAN 1 1 2016

**In the United States Patent and Trademark Office**

Appn. Number:    14/028,786
Appn. Filed       09/17/2013
Applicants:       Roy Lique
Title:            Digital Camera Lens Guard and Use Extender
Examiner/GAU:     Christopher Mahoney/Art Unit 2852

Chino Hills, January 6, 2016

**AMENDMENT B**

Mail Stop Non-Fee Amendments
Commissioner for Patents
P.O. Box 1450
Alexandria, VA 22313-1450

Sir:

In response to the Office Action Mailed November 4, 2015, please amend the above application as follows:

CLAIMS: Amendments to the claims begin on page 2 of this Amendment
REMARKS: Remarks begin on page 3 of this Amendment.

## Issue Classification – Page 1

| *Issue Classification* | Application/Control No. | Applicant(s)/Patent Under Reexamination |
|---|---|---|
| | 14028786 | LIQUE, ROY |
| | Examiner | Art Unit |
| | CHRIST MAHONEY | 2852 |

**CPC**

| Symbol | | | | Type | Version |
|---|---|---|---|---|---|
| H04N | 5 | 2252 | | F | 2013-01-01 |
| H04N | 5 | 2254 | | I | 2013-01-01 |
| G03B | 17 | 12 | | A | 2013-01-01 |
| | | | | | |
| | | | | | |
| | | | | | |
| | | | | | |
| | | | | | |
| | | | | | |
| | | | | | |
| | | | | | |
| | | | | | |
| | | | | | |
| | | | | | |

**CPC Combination Sets**

| Symbol | | | Type | Set | Ranking | Version |
|---|---|---|---|---|---|---|
| | | | | | | |
| | | | | | | |

| NONE | | Total Claims Allowed: |
|---|---|---|
| | | 3 |
| (Assistant Examiner) | (Date) | |
| /Christopher Mahoney/ Primary Examiner Art Unit 2852 | 02/01/2016 | O.G. Print Claim(s) / O.G. Print Figure |
| (Primary Examiner) | (Date) | 1 / 1 |

U.S. Patent and Trademark Office

Part of Paper No. 42

## Issue Classification – Page 2

| *Issue Classification* | Application/Control No. | Applicant(s)/Patent Under Reexamination |
|---|---|---|
| | 14028786 | LIQUE, ROY |
| | Examiner | Art Unit |
| | CHRIST MAHONEY | 2852 |

| US ORIGINAL CLASSIFICATION | | INTERNATIONAL CLASSIFICATION | |
|---|---|---|---|
| CLASS | SUBCLASS | CLAIMED | NON-CLAIMED |
| | | H 0 4 N | S - 22/1 (2008.01 01) | |

| CROSS REFERENCE(S) |
|---|
| CLASS | SUBCLASS (ONE SUBCLASS PER BLOCK) |

| NONE | | Total Claims Allowed: | |
|---|---|---|---|
| (Assistant Examiner) | (Date) | 3 | |
| /Christopher Mahoney/ Primary Examiner Art Unit 2852 | 02/01/2016 | O.G. Print Claim(s) | O.G. Print Figure |
| (Primary Examiner) | (Date) | 1 | 1 |

U.S. Patent and Trademark Office

Part of Paper No. 02

## Issue Classification – Page 3

| *Issue Classification* | Application/Control No. | Applicant(s)/Patent Under Reexamination |
|---|---|---|
| | 14028786 | LIQUE, ROY |
| (barcode) | Examiner | Art Unit |
| | CHRIST MAHONEY | 2852 |

☑ Claims renumbered in the same order as presented by applicant  ☐ CPA  ☐ T.D.  ☐ R.1.47

| Final | Original | Final | Original | Final | Original | Final | Original | Final | Original | Final | Original | Final | Original | Final | Original |
|---|---|---|---|---|---|---|---|---|---|---|---|---|---|---|---|
| | | | | | | | | | | | | | | | |
| | | | | | | | | | | | | | | | |
| | | | | | | | | | | | | | | | |
| | | | | | | | | | | | | | | | |
| | | | | | | | | | | | | | | | |
| | | | | | | | | | | | | | | | |
| | | | | | | | | | | | | | | | |
| | | | | | | | | | | | | | | | |
| | | | | | | | | | | | | | | | |
| | | | | | | | | | | | | | | | |
| | | | | | | | | | | | | | | | |
| | | | | | | | | | | | | | | | |
| | | | | | | | | | | | | | | | |

NONE

| | | | Total Claims Allowed: |
|---|---|---|---|
| (Assistant Examiner) | | (Date) | 3 |
| /Christopher Mahoney/ Primary Examiner Art Unit 2852 | | 02/01/2016 | O.G. Print Claim(s) / O.G. Print Figure |
| (Primary Examiner) | | (Date) | 1 / 1 |

U.S. Patent and Trademark Office

Part of Paper No. 02

# Search Notes

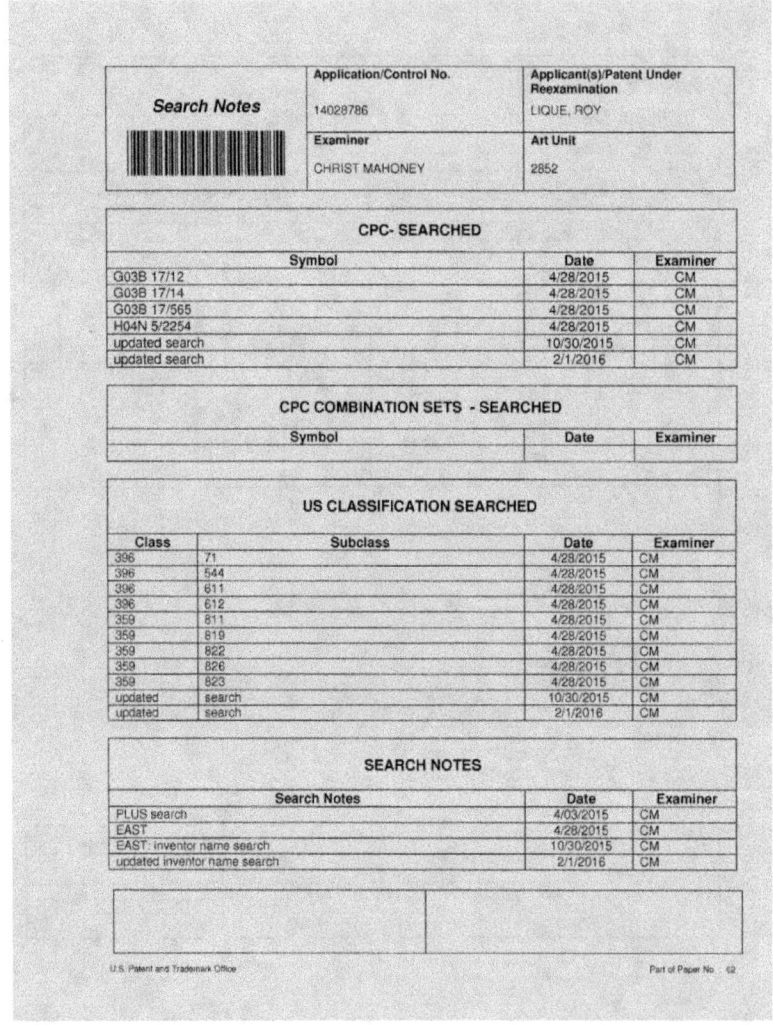

| Search Notes | Application/Control No. 14028786 | Applicant(s)/Patent Under Reexamination LIQUE, ROY |
|---|---|---|
| | Examiner CHRIST MAHONEY | Art Unit 2852 |

### CPC- SEARCHED

| Symbol | Date | Examiner |
|---|---|---|
| G03B 17/12 | 4/28/2015 | CM |
| G03B 17/14 | 4/28/2015 | CM |
| G03B 17/565 | 4/28/2015 | CM |
| H04N 5/2254 | 4/28/2015 | CM |
| updated search | 10/30/2015 | CM |
| updated search | 2/1/2016 | CM |

### CPC COMBINATION SETS - SEARCHED

| Symbol | Date | Examiner |
|---|---|---|
| | | |

### US CLASSIFICATION SEARCHED

| Class | Subclass | Date | Examiner |
|---|---|---|---|
| 396 | 71 | 4/28/2015 | CM |
| 396 | 544 | 4/28/2015 | CM |
| 396 | 611 | 4/28/2015 | CM |
| 396 | 612 | 4/28/2015 | CM |
| 359 | 811 | 4/28/2015 | CM |
| 359 | 819 | 4/28/2015 | CM |
| 359 | 822 | 4/28/2015 | CM |
| 359 | 826 | 4/28/2015 | CM |
| 359 | 823 | 4/28/2015 | CM |
| updated | search | 10/30/2015 | CM |
| updated | search | 2/1/2016 | CM |

### SEARCH NOTES

| Search Notes | Date | Examiner |
|---|---|---|
| PLUS search | 4/03/2015 | CM |
| EAST | 4/28/2015 | CM |
| EAST: inventor name search | 10/30/2015 | CM |
| updated inventor name search | 2/1/2016 | CM |

U.S. Patent and Trademark Office

Part of Paper No. 62

**Interference Search**

| INTERFERENCE SEARCH | | | |
|---|---|---|---|
| US Class/ CPC Symbol | US Subclass / CPC Group | Date | Examiner |
| G03B | 17/12 | 10/30/2015 | CM |
| G03B | 17/14 | 10/30/2015 | CM |
| G03B | 17/565 | 10/30/2015 | CM |
| H04N | 5/2254 | 10/30/2015 | CM |
| 396 | 71 | 10/30/2015 | CM |
| 396 | 544 | 10/30/2015 | CM |
| 396 | 611 | 10/30/2015 | CM |
| 396 | 612 | 10/30/2015 | CM |
| 359 | 811 | 10/30/2015 | CM |
| 359 | 819 | 10/30/2015 | CM |
| 359 | 822 | 10/30/2015 | CM |
| 359 | 826 | 10/30/2015 | CM |
| 359 | 823 | 10/30/2015 | CM |
| updated | search | 2/1/2016 | CM |

## Notice of Allowance and Fees Due
Maintenance Fees

All utility patents that issue from applications filed on or after December 12, 1980 are subject to the payment of maintenance fees which must be paid to maintain the patent in force. These fees are due at 3.5, 7.5 and 11.5 years from the date the patent is granted and can be paid without a surcharge during the "window period," which is the six-month period preceding each due date, e.g., three years to three years and six months. (See fee schedule for a list of maintenance fees.) In submitting maintenance fees and any necessary surcharges, identification of the patents for which maintenance fees are being paid must include the patent number, and the application number of the U.S. application for the patent on which the maintenance fee is being paid. If the payment includes identification of only the patent number, the Office may apply payment to the patent identified by patent number in the payment or the Office may return the payment. (See 37, Code of Federal Regulations, section 1.366(c).)

Failure to pay the current maintenance fee on time may result in expiration of the patent. A six-month grace period is provided when the maintenance fee may be paid with a surcharge. The grace period is the six-month period immediately following the due date. The USPTO does not mail notices to patent owners that maintenance fees are due. If, however, the maintenance fee is not paid on time, efforts are made to remind the responsible party that the maintenance fee may be paid during the grace period with a surcharge. If the maintenance fee is not paid on time and the maintenance fee and surcharge are not paid during the grace period, the patent expires on the date the grace period ends.

# Notice of Allowance and Fees Due – Page 1

UNITED STATES PATENT AND TRADEMARK OFFICE

UNITED STATES DEPARTMENT OF COMMERCE
United States Patent and Trademark Office
Address: COMMISSIONER FOR PATENTS
P.O. Box 1450
Alexandria, Virginia 22313-1450
www.uspto.gov

## NOTICE OF ALLOWANCE AND FEE(S) DUE

Roy Lique
2650 Lookout Cir.
Chino Hills, CA 91709

| EXAMINER |
|---|
| MAHONEY, CHRISTOPHER E |

| ART UNIT | PAPER NUMBER |
|---|---|
| 2852 | |

DATE MAILED: 02/08/2016

| APPLICATION NO. | FILING DATE | FIRST NAMED INVENTOR | ATTORNEY DOCKET NO. | CONFIRMATION NO. |
|---|---|---|---|---|
| 14/028,786 | 09/17/2013 | Roy Lique | Camera Coup | 6716 |

TITLE OF INVENTION: Digital Camera Lens Guard and Use Extender

| APPLN. TYPE | ENTITY STATUS | ISSUE FEE DUE | PUBLICATION FEE DUE | PREV. PAID ISSUE FEE | TOTAL FEE(S) DUE | DATE DUE |
|---|---|---|---|---|---|---|
| nonprovisional | MICRO | $240 | $0 | $0 | $240 | 05/09/2016 |

THE APPLICATION IDENTIFIED ABOVE HAS BEEN EXAMINED AND IS ALLOWED FOR ISSUANCE AS A PATENT. **PROSECUTION ON THE MERITS IS CLOSED.** THIS NOTICE OF ALLOWANCE IS NOT A GRANT OF PATENT RIGHTS. THIS APPLICATION IS SUBJECT TO WITHDRAWAL FROM ISSUE AT THE INITIATIVE OF THE OFFICE OR UPON PETITION BY THE APPLICANT. SEE 37 CFR 1.313 AND MPEP 1308.

THE ISSUE FEE AND PUBLICATION FEE (IF REQUIRED) MUST BE PAID WITHIN **THREE MONTHS** FROM THE MAILING DATE OF THIS NOTICE OR THIS APPLICATION SHALL BE REGARDED AS ABANDONED. **THIS STATUTORY PERIOD CANNOT BE EXTENDED.** SEE 35 U.S.C. 151. THE ISSUE FEE DUE INDICATED ABOVE DOES NOT REFLECT A CREDIT FOR ANY PREVIOUSLY PAID ISSUE FEE IN THIS APPLICATION. IF AN ISSUE FEE HAS PREVIOUSLY BEEN PAID IN THIS APPLICATION (AS SHOWN ABOVE), THE RETURN OF PART B OF THIS FORM WILL BE CONSIDERED A REQUEST TO REAPPLY THE PREVIOUSLY PAID ISSUE FEE TOWARD THE ISSUE FEE NOW DUE.

**HOW TO REPLY TO THIS NOTICE:**

I. Review the ENTITY STATUS shown above. If the ENTITY STATUS is shown as SMALL or MICRO, verify whether entitlement to that entity status still applies.

If the ENTITY STATUS is the same as shown above, pay the TOTAL FEE(S) DUE shown above.

If the ENTITY STATUS is changed from that shown above, on PART B - FEE(S) TRANSMITTAL, complete section number 5 titled "Change in Entity Status (from status indicated above)".

For purposes of this notice, small entity fees are 1/2 the amount of undiscounted fees, and micro entity fees are 1/2 the amount of small entity fees.

II. PART B - FEE(S) TRANSMITTAL, or its equivalent, must be completed and returned to the United States Patent and Trademark Office (USPTO) with your ISSUE FEE and PUBLICATION FEE (if required). If you are charging the fee(s) to your deposit account, section "4b" of Part B - Fee(s) Transmittal should be completed and an extra copy of the form should be submitted. If an equivalent of Part B is filed, a request to reapply a previously paid issue fee must be clearly made, and delays in processing may occur due to the difficulty in recognizing the paper as an equivalent of Part B.

III. All communications regarding this application must give the application number. Please direct all communications prior to issuance to Mail Stop ISSUE FEE unless advised to the contrary.

**IMPORTANT REMINDER: Utility patents issuing on applications filed on or after Dec. 12, 1980 may require payment of maintenance fees. It is patentee's responsibility to ensure timely payment of maintenance fees when due.**

Page 1 of 3

PTOL-85 (Rev. 02/11)

# Notice of Allowance and Fees Due – Page 2

**PART B - FEE(S) TRANSMITTAL**

Complete and send this form, together with applicable fee(s), to: **Mail**    Mail Stop ISSUE FEE
Commissioner for Patents
P.O. Box 1450
Alexandria, Virginia 22313-1450
or **Fax**   (571)-273-2885

INSTRUCTIONS: This form should be used for transmitting the ISSUE FEE and PUBLICATION FEE (if required). Blocks 1 through 5 should be completed where appropriate. All further correspondence including the Patent, advance orders and notification of maintenance fees will be mailed to the current correspondence address as indicated unless corrected below or directed otherwise in Block 1, by (a) specifying a new correspondence address; and/or (b) indicating a separate "FEE ADDRESS" for maintenance fee notifications.

CURRENT CORRESPONDENCE ADDRESS (Note: Use Block 1 for any change of address)

| | | |
|---|---|---|
| 117033 | 7590 | 02/06/2016 |

Roy Lique
2650 Lookout Cir.
Chino Hills, CA 91709

Note: A certificate of mailing can only be used for domestic mailings of the Fee(s) Transmittal. This certificate cannot be used for any other accompanying papers. Each additional paper, such as an assignment or formal drawing, must have its own certificate of mailing or transmission.

**Certificate of Mailing or Transmission**
I hereby certify that this Fee(s) Transmittal is being deposited with the United States Postal Service with sufficient postage for first class mail in an envelope addressed to the Mail Stop ISSUE FEE address above, or being facsimile transmitted to the USPTO (571) 273-2885, on the date indicated below.

ROY LIQUE _____ (Depositor's name)
Roylique _____ (Signature)
02-12-2016 _____ (Date)

| APPLICATION NO. | FILING DATE | FIRST NAMED INVENTOR | ATTORNEY DOCKET NO. | CONFIRMATION NO. |
|---|---|---|---|---|
| 14/028,786 | 09/17/2013 | Roy Lique | Camera Coup | 6716 |

TITLE OF INVENTION: Digital Camera Lens Guard and Use Extender

| APPLN. TYPE | ENTITY STATUS | ISSUE FEE DUE | PUBLICATION FEE DUE | PREV. PAID ISSUE FEE | TOTAL FEE(S) DUE | DATE DUE |
|---|---|---|---|---|---|---|
| nonprovisional | MICRO | $240 | $0 | $0 | $240 | 05/09/2016 |

| EXAMINER | ART UNIT | CLASS-SUBCLASS |
|---|---|---|
| MAHONEY, CHRISTOPHER E | 2852 | 396-529000 |

1. Change of correspondence address or indication of "Fee Address" (37 CFR 1.363).
☐ Change of correspondence address (or Change of Correspondence Address form PTO/SB/122) attached.
☐ "Fee Address" indication (or "Fee Address" Indication form PTO/SB/47; Rev 03-02 or more recent) attached. **Use of a Customer Number is required.**

2. For printing on the patent front page, list
(1) The names of up to 3 registered patent attorneys or agents OR, alternatively     1 _____
(2) The name of a single firm (having as a member a registered attorney or agent) and the names of up to 2 registered patent attorneys or agents. If no name is listed, no name will be printed.     2 _____     3 _____

3. ASSIGNEE NAME AND RESIDENCE DATA TO BE PRINTED ON THE PATENT (print or type)

PLEASE NOTE: Unless an assignee is identified below, no assignee data will appear on the patent. If an assignee is identified below, the document has been filed for recordation as set forth in 37 CFR 3.11. Completion of this form is NOT a substitute for filing an assignment.

(A) NAME OF ASSIGNEE                (B) RESIDENCE: (CITY and STATE OR COUNTRY)

Please check the appropriate assignee category or categories (will not be printed on the patent):   ☐ Individual  ☐ Corporation or other private group entity  ☐ Government

4a. The following fee(s) are submitted:
☑ Issue Fee
☐ Publication Fee (No small entity discount permitted)
☐ Advance Order - # of Copies _____

4b. Payment of Fee(s): (Please first reapply any previously paid issue fee shown above)
☑ A check is enclosed.
☐ Payment by credit card. Form PTO-2038 is attached.
☐ The director is hereby authorized to charge the required fee(s), any deficiency, or credits any overpayment, to Deposit Account Number _____ (enclose an extra copy of this form).

5. Change in Entity Status (from status indicated above)
☐ Applicant certifying micro entity status. See 37 CFR 1.29
☐ Applicant asserting small entity status. See 37 CFR 1.27
☐ Applicant changing to regular undiscounted fee status.

NOTE: Absent a valid certification of Micro Entity Status (see forms PTO/SB/15A and 15B), issue fee payment in the micro entity amount will not be accepted at the risk of application abandonment.
NOTE: If the application was previously under micro entity status, checking this box will be taken to be a notification of loss of entitlement to micro entity status.
NOTE: Checking this box will be taken to be a notification of loss of entitlement to small or micro entity status, as applicable.

NOTE: This form must be signed in accordance with 37 CFR 1.31 and 1.33. See 37 CFR 1.4 for signature requirements and certifications.

Authorized Signature   Roylique _____   Date   02-12-2016 _____
Typed or printed name   Roy LIQUE _____   Registration No.   117033 _____

Page 2 of 3

PTOL-85 Part B (10-13) Approved for use through 10/31/2013.     OMB 0651-0033     U.S. Patent and Trademark Office; U.S. DEPARTMENT OF COMMERCE

## Determination of Patent Term Adjustment – Page 1

UNITED STATES PATENT AND TRADEMARK OFFICE

UNITED STATES DEPARTMENT OF COMMERCE
United States Patent and Trademark Office
Address: COMMISSIONER FOR PATENTS
P.O. Box 1450
Alexandria, Virginia 22313-1450
www.uspto.gov

| APPLICATION NO. | FILING DATE | FIRST NAMED INVENTOR | ATTORNEY DOCKET NO. | CONFIRMATION NO. |
|---|---|---|---|---|
| 14/028,786 | 09/17/2013 | Roy Lique | Camera Coup | 6716 |

| | |
|---|---|
| 13703    7506    02/08/2016 | EXAMINER |
| Roy Lique | MAHONEY, CHRISTOPHER E |
| 2650 Lookout Cir. | |
| Chino Hills, CA 91709 | ART UNIT    PAPER NUMBER |
| | 2852 |

DATE MAILED: 02/08/2016

**Determination of Patent Term Adjustment under 35 U.S.C. 154 (b)**
(Applications filed on or after May 29, 2000)

The Office has discontinued providing a Patent Term Adjustment (PTA) calculation with the Notice of Allowance.

Section 1(h)(2) of the AIA Technical Corrections Act amended 35 U.S.C. 154(b)(3)(B)(i) to eliminate the requirement that the Office provide a patent term adjustment determination with the notice of allowance. See Revisions to Patent Term Adjustment, 78 Fed. Reg. 19416, 19417 (Apr. 1, 2013). Therefore, the Office is no longer providing an initial patent term adjustment determination with the notice of allowance. The Office will continue to provide a patent term adjustment determination with the Issue Notification Letter that is mailed to applicant approximately three weeks prior to the issue date of the patent, and will include the patent term adjustment on the patent. Any request for reconsideration of the patent term adjustment determination (or reinstatement of patent term adjustment) should follow the process outlined in 37 CFR 1.705.

Any questions regarding the Patent Term Extension or Adjustment determination should be directed to the Office of Patent Legal Administration at (571)-272-7702. Questions relating to issue and publication fee payments should be directed to the Customer Service Center of the Office of Patent Publication at 1-(888)-786-0101 or (571)-272-4200.

Page 3 of 3

PTOL-85 (Rev. 02/11)

# Determination of Patent Term Adjustment – Page 2

### OMB Clearance and PRA Burden Statement for PTOL-85 Part B

The Paperwork Reduction Act (PRA) of 1995 requires Federal agencies to obtain Office of Management and Budget approval before requesting most types of information from the public. When OMB approves an agency request to collect information from the public, OMB (i) provides a valid OMB Control Number and expiration date for the agency to display on the instrument that will be used to collect the information and (ii) requires the agency to inform the public about the OMB Control Number's legal significance in accordance with 5 CFR 1320.5(b).

The information collected by PTOL-85 Part B is required by 37 CFR 1.311. The information is required to obtain or retain a benefit by the public which is to file (and by the USPTO to process) an application. Confidentiality is governed by 35 U.S.C. 122 and 37 CFR 1.14. This collection is estimated to take 12 minutes to complete, including gathering, preparing, and submitting the completed application form to the USPTO. Time will vary depending upon the individual case. Any comments on the amount of time you require to complete this form and/or suggestions for reducing this burden, should be sent to the Chief Information Officer, U.S. Patent and Trademark Office, U.S. Department of Commerce, P.O. Box 1450, Alexandria, Virginia 22313-1450. DO NOT SEND FEES OR COMPLETED FORMS TO THIS ADDRESS. SEND TO: Commissioner for Patents, P.O. Box 1450, Alexandria, Virginia 22313-1450. Under the Paperwork Reduction Act of 1995, no persons are required to respond to a collection of information unless it displays a valid OMB control number.

### Privacy Act Statement

**The Privacy Act of 1974 (P.L. 93-579)** requires that you be given certain information in connection with your submission of the attached form related to a patent application or patent. Accordingly, pursuant to the requirements of the Act, please be advised that: (1) the general authority for the collection of this information is 35 U.S.C. 2(b)(2); (2) furnishing of the information solicited is voluntary; and (3) the principal purpose for which the information is used by the U.S. Patent and Trademark Office is to process and/or examine your submission related to a patent application or patent. If you do not furnish the requested information, the U.S. Patent and Trademark Office may not be able to process and/or examine your submission, which may result in termination of proceedings or abandonment of the application or expiration of the patent.

The information provided by you in this form will be subject to the following routine uses:

1. The information on this form will be treated confidentially to the extent allowed under the Freedom of Information Act (5 U.S.C. 552) and the Privacy Act (5 U.S.C 552a). Records from this system of records may be disclosed to the Department of Justice to determine whether disclosure of these records is required by the Freedom of Information Act.
2. A record from this system of records may be disclosed, as a routine use, in the course of presenting evidence to a court, magistrate, or administrative tribunal, including disclosures to opposing counsel in the course of settlement negotiations.
3. A record in this system of records may be disclosed, as a routine use, to a Member of Congress submitting a request involving an individual, to whom the record pertains, when the individual has requested assistance from the Member with respect to the subject matter of the record.
4. A record in this system of records may be disclosed, as a routine use, to a contractor of the Agency having need for the information in order to perform a contract. Recipients of information shall be required to comply with the requirements of the Privacy Act of 1974, as amended, pursuant to 5 U.S.C. 552a(m).
5. A record related to an International Application filed under the Patent Cooperation Treaty in this system of records may be disclosed, as a routine use, to the International Bureau of the World Intellectual Property Organization, pursuant to the Patent Cooperation Treaty.
6. A record in this system of records may be disclosed, as a routine use, to another federal agency for purposes of National Security review (35 U.S.C. 181) and for review pursuant to the Atomic Energy Act (42 U.S.C. 218(c)).
7. A record from this system of records may be disclosed, as a routine use, to the Administrator, General Services, or his/her designee, during an inspection of records conducted by GSA as part of that agency's responsibility to recommend improvements in records management practices and programs, under authority of 44 U.S.C. 2904 and 2906. Such disclosure shall be made in accordance with the GSA regulations governing inspection of records for this purpose, and any other relevant (i.e. GSA or Commerce) directive. Such disclosure shall not be used to make determinations about individuals.
8. A record from this system of records may be disclosed, as a routine use, to the public after either publication of the application pursuant to 35 U.S.C. 122(b) or issuance of a patent pursuant to 35 U.S.C. 151. Further, a record may be disclosed, subject to the limitations of 37 CFR 1.14, as a routine use, to the public if the record was filed in an application which became abandoned or in which the proceedings were terminated and which application is referenced by either a published application, an application open to public inspection or an issued patent.
9. A record from this system of records may be disclosed, as a routine use, to a Federal, State, or local law enforcement agency, if the USPTO becomes aware of a violation or potential violation of law or regulation.

**Fee enclosed – patent granted!**

| | | | |
|---|---|---|---|
| PLEASE POST THIS PAYMENT FOR OUR MUTUAL CUSTOMER | | | |
| Account: **APPLICATION NO. 14/028,786** | | | $240.00 |

Please Direct Any Questions To
(855) 739-0856
Payment Processing Center
P.O. Box 1029
Hickory, NC 28603-1029

70-2382/719

**0041107405**

ROY LIQUE
2650 LOOKOUT CIR
CHINO HILLS, CA 91709-1193

NORTHERN TRUST

**February 11, 2016**

ty TWO HUNDRED FORTY AND 00/100 ------------------------------------------------ DOLLARS

$ *******240.00

TO
THE
ORDER
OF

DIRECTOR OF THE US PATENT OFFICE
C/O ROY LIQUE
2650 LOOKOUT CIR
CHINO HILLS, CA 91709-1193

REMITTANCE VOID IF NOT CASHED WITHIN 90 DAYS

AUTHORIZED SIGNATURE

⑈004110740 5⑈ ⑆071923828⑆ 003511000 3⑈

# Part 8 – Index

www.ingramcontent.com/pod-product-compliance
Lightning Source LLC
Chambersburg PA
CBHW081142180526
45170CB00006B/1896